FINDINGS ON LIGHT

Curated and edited by
Hester Aardse, Astrid Alben

PARS Foundation
Lars Müller Publishers

PARS is an arts and sciences organization led by art historian Hester Aardse and poet Astrid Alben. We invite artists and scientists to share their most revealing, beautiful, funny and mind-boggling research with us around particular topics. PARS publishes these in a publication series, such as *Findings on Ice* (2007) and *Findings on Elasticity* (2010). PARS also curates events, which are a mixture of art, theatre and scientific experiment, at different locations. Our aim is to stimulate curiosity and to celebrate beauty in knowledge.

www.parsfoundation.com

Dear Reader,

Light is one of the most essential elements for the existence of life on earth. Plants, animals and humans depend on it. It is the strongest and fastest form of energy. Our biorhythm is based on light; it regulates our biochemical and neurological systems, but living in cities generally implies losing track of it. Without light we get depressed and die from lack of vitamins created by the workings of light. However, too much light might overwork us as well. Yet some deepsea animals exist without any light at all, in utter darkness. Without light we cannot see or distinguish shapes and colours. Yet we don't necessarily need our eyes to perceive it.

So what is it when we see light, how do we see it, and what can it do? And sometimes the light at the end of a tunnel really *is* a train.

Findings on Light features the work of artists and scientists whose binding interest is light. Light is one of the most common yet mysterious phenomena in our universe and it has captivated creative thinkers for millennia. In the process of making this book we discovered it is even more intertwined with everything in and around us, and far more elusive than we initially thought. Even though light is all around us, no one fully understands it yet.

We invited the contributors, with the help of advisors from different fields of expertise, and stipulated only two conditions: that their findings relate to light, and that it reflect the language of their profession. As a result *Findings on Light* is a book about light but equally it is a book about the ways in which artists and scientists describe and engage with the world.

Their findings range from the quirky, humorous and beautiful, to the mind-bogglingly complex and disturbing. While compiling this book we were astounded by the sheer inexhaustible amount of ideas, research and beauty we encountered. As the material poured in we were moved, made to laugh, to think, to discuss, and raised our eyebrows and scratched our heads at complexities. This is not a book that sets out to *explain* light but it will take you on an *exploration* of light: from solar energy that can replace kerosene, to laser therapy that corrects our eyesight, to light that transports bundles of information or is bent to camouflage objects, to the development of new medication and technologies through research into bioluminescence; to ever-larger telescopes allowing us to explore our universe, bursting at the seams; and to the music we are surrounded by, and the paintings, stories and installations that help us anchor our emotions and keep track of our culture through the ages.

It is artists and scientists who shape the way we look at the world. What binds them is their curiosity and passion, and it is this curiosity and passion to keep reinventing and reimagining the world that we want to share with you, through the topic of light. As with *Findings on Ice* and *Findings on Elasticity* there is much for you to discover and marvel at. We hope you will enjoy dipping in and out of these findings on light as much as we did, and do your own reimagining of the world with it.

Hester Aardse and Astrid Alben
The Editors

Adam Fuss, photographer

UNTITLED

ABYSS HYPNOTIZING

Unique cibachrome photogram, 1992
162.6 × 125.7 centimetres

Hans Kristian Eriksen, astrophysicist

BURSTING AT THE SEAMS

BIG BANG | ORIGIN | UNIVERSE | RHYTHM | DIRECTION

Some three thousand years ago, the creation of the universe was succinctly summarized by the authors of Genesis as, "And God said, 'Let there be light': and there was light." Fast-forward a few thousand years, and we are finally able to add an astonishing amount of detail to this sentence, as we aspire to measure the echoes of spacetime bending under the strain of quantum fluctuations when it was only 0.000 000 000 000 000 000 000 000 000 000 001 seconds old. Yet, even though the various physical processes acting in the newborn universe are remarkably well understood, the most basic and fundamental questions remain as mysterious and unapproachable as they did thousands of years ago: what are physical laws, and why does the universe obey them?

For all practical purposes, scientific physical cosmology started with Einstein's Theory of General Relativity (GR), published in 1916. This theory can be summarized in the iconic Einstein field equation,

$$E_{\mu\nu} = 8\pi\, G T_{\mu\nu}$$

where the left-hand side describes the geometry of space, and the right-hand side describes the matter contents of space (stars, galaxies, dark matter and energy). In the words of Professor John Archibald Wheeler, this equation states that "space tells matter how to move; matter tells space how to curve", and it applies equally well to apples falling from trees and the Earth moving around the sun, as to the universe as a whole.

When applied to the universe, however, this equation makes a dramatic prediction: it doesn't allow for stable, static solutions. Things have to move and evolve. If one tries to set up an initially static universe in which no galaxy moves with respect to others, then gravity will make things implode, much like an apple will fall to the ground if left alone. The only way to avoid this implosion is if space itself expands, such that all galaxies on average move away from all others. The American astronomer Edwin Hubble discovered such expansion in 1929, forever changing our view of the universe.

Universal expansion has several profound consequences. For instance, if the observable universe will be larger tomorrow than it is today, it must have been smaller yesterday. And even smaller the day before that. In fact, if one keeps going backward in time, one will find that at some point the entire observable universe was contained within a very small volume indeed – a billiard ball, a marble, or even the nucleus of an atom. And according to GR and the most recent cosmological measurements, that happened 13.8 billion years ago, with an uncertainty of 50 million years.

However, it is well known that if one compresses a gas into a smaller volume, it heats up. To experience this in daily life, simply inflate a bicycle tyre with a hand-driven air pump; the pump becomes warm from the increased air pressure. The same happened in the early universe. As we go back in time, matter density increases, and therefore the pressure and temperature also increases. Indeed, if one looks far enough into the past, temperatures reach such heights as to burn everything to pieces. The extreme photon intensity rips all

01   Cosmic Microwave Background intensity: baby picture of the early universe, as measured by Planck, a European-funded ESA satellite mission. The red and blue spots indicate regions that happened to be slightly warmer or colder than average when the universe was 380 000 years old. Galaxies later formed most easily in the blue regions.
02   Galactic dust polarization: amplitude of polarized light from vibrating dust grains in the Milky Way
03   The measurement of the polarized Cosmic Microwave Background sky as observed by Planck.
04   Scalar simulation: a perfect simulation including only scalar (normal density) variations.
05   Tensor simulation: a perfect simulation including only tensor (inflationary gravity wave) variations.
06   A polarization of galactic dust.

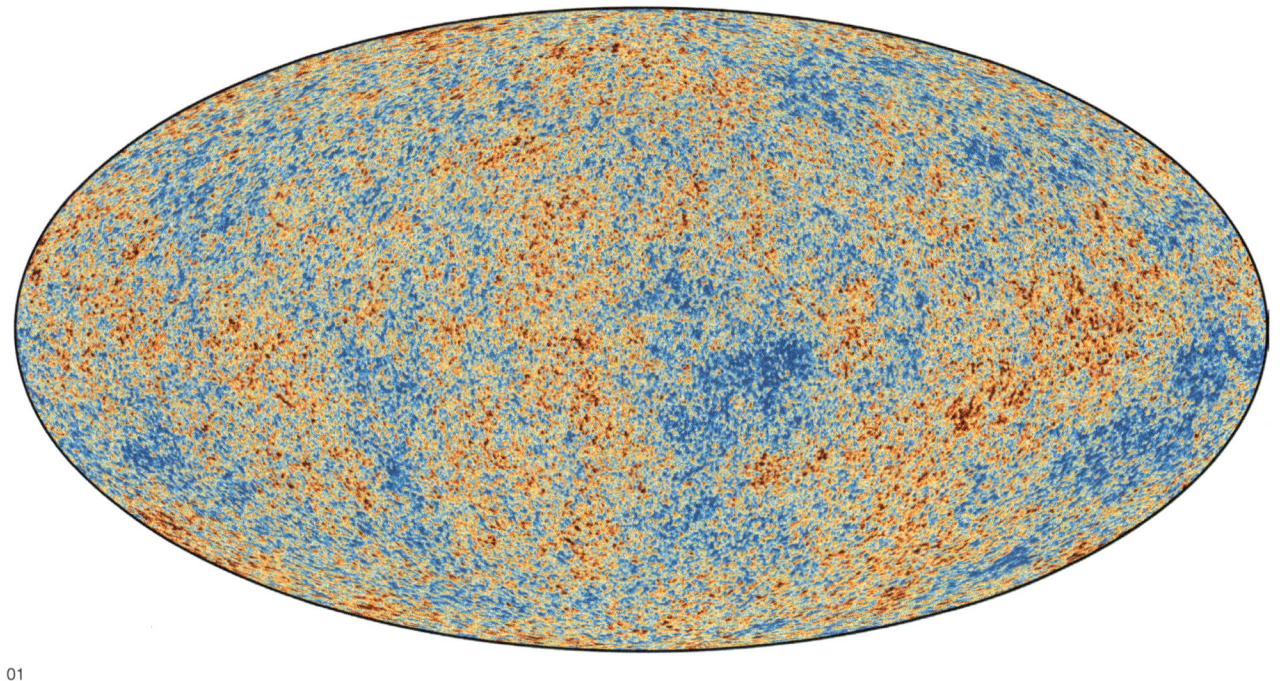

01

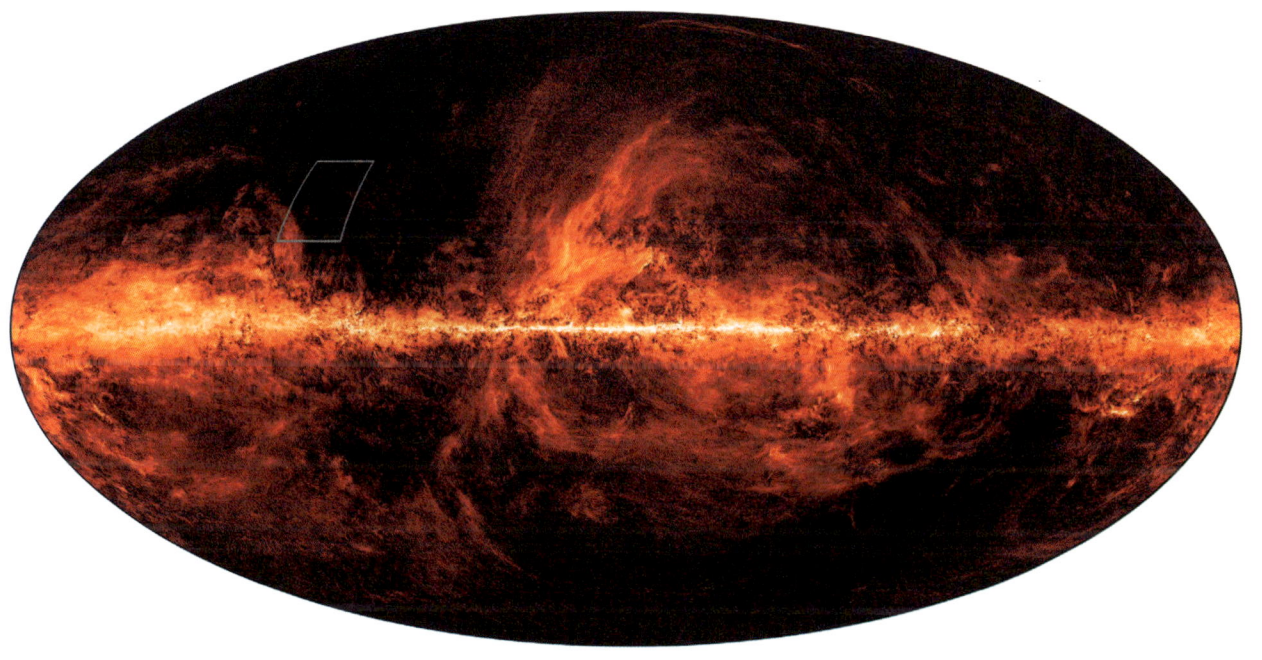

02

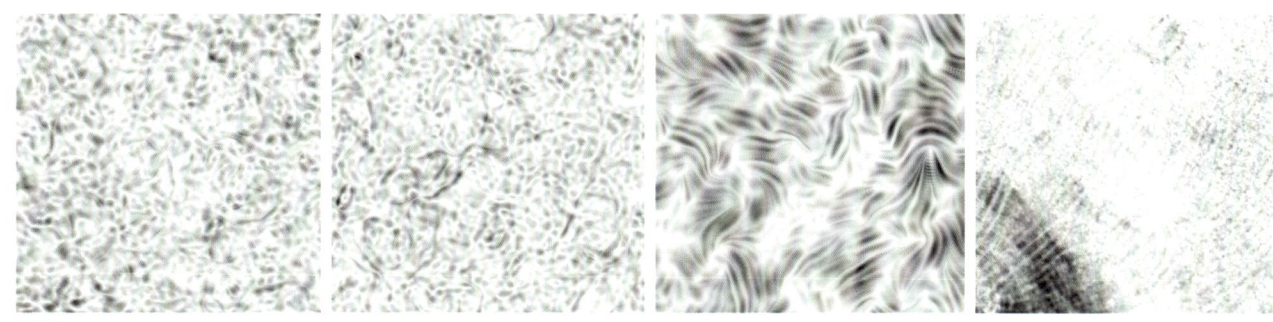

03-06

#Time (sec)	Voltage (det1)	Voltage (det2)	Voltage	(det3)

compound objects apart: molecules, heavy atoms and even hydrogen atoms. In the beginning, the only things that could exist were particles that cannot be further subdivided – elementary particles. The universe was indeed created with a burst of light, and today this miracle is known as the Big Bang.

Luckily for us, it is still possible to observe the by-products of the processes that acted in the universe during the first few hundred thousand years after the Big Bang, in the form of the Cosmic Microwave Background (CMB). Reversing the process once again and now going forward in time, starting from the Big Bang, the universe cooled rapidly as it expanded. During the first 380 000 years the universe was so hot that any neutral atoms that might have formed immediately were torn to pieces by the intense light. The photons in turn constantly collided with free electrons in a process called Thomson scattering, and they could not move more than a metre or two before changing direction; the light was firmly trapped within a dense electron gas. If it was possible to step into that gas, it would be exactly like walking through a dense fog. You would see your own hand, but not the person standing three metres away from you.

This changed when the temperature fell below 3000°K (or about 2700°C), and the average photon energy decreased below 13.6 eV, the binding energy of hydrogen. At this point, electrons and protons rapidly combined into neutral hydrogen. And without any free electrons left in the universe, there was nothing for the light to scatter on any more. Suddenly, the photons were free to travel through the entire universe. Since that time, the universe has expanded further by roughly a factor of 1100, and the temperature has continued decreasing by the same factor. The photons that were originally released with a temperature of 3000°K (comparable to that of a cool star), are still racing through the universe, but now with an average temperature of 2.7°K. Those same photons may therefore be detected and measured using ultra-sensitive radio antennas tuned to a frequency of 30–300 GHz. This cool background light is what is today known as the Cosmic Microwave Background, and scientists all over the world spend billions of euros to measure it with exquisite precision.

The most accurate CMB measurements to date have been produced by Planck, a European satellite mission launched in 2009. The final CMB temperature map from Planck is shown in [01]. The red and blue spots correspond to directions in which the photon temperature was slightly hotter or cooler than average, because the matter density happened to be slightly lower or higher than the average. The blue spots in this map were the seeds of later galaxy formation; most likely, Earth itself originated in one such cold spot. This image therefore represents a baby picture of the universe, and as such contains a vast amount of cosmological information. When was the universe created? How much matter does it contain? When did the first stars ignite? This map is where all those questions find their answers, and already two Nobel Prizes have been awarded based on those types of questions; first to Robert Wilson and Arnos Penzias in 1977 for the discovery of the CMB (that is, the very existence of the 2.7°K background), and then to John Mather and George Smoot for the discovery of the first fluctuations around this average, as seen in [01].

However, while understanding the CMB intensity has proved a milestone in human achievement, light contains more information than just intensity. It may also exhibit a preferred direction, called polarization. The intensity may be slightly stronger along one plane than along the orthogonal plane. In daily life, this phenomenon is perhaps best observed through polarized sunglasses, which filter light that is scattered off (and therefore parallel to) the ground. Likewise, the CMB may also be polarized because Thomson scattering with electrons is an intrinsically polarized process. However, in order to generate a net polarization as observed from a distance, the intensity along one direction must be higher than along the orthogonal direction; it must be warmer above and below a electron compared to its left and right. This can be achieved by two different mechanisms: first, the same density variations that create the red and blue spots in [01] may by chance also create the required temperature alignment. Such variations are called *scalar* perturbations, and originate from the same time in history as when the photons were originally released, some 380 000 years after the Big Bang. However, there is a second mechanism that can also create similar patterns, namely gravity waves, originally predicted by Einstein himself. Because space can bend, it is possible to squeeze it in one direction and expand it in the orthogonal direction. If this happens in the early universe, photons would heat up in the squeezed direction, and cool off in the expanded direction, resulting in a net polarization. This type of polarization is called *tensor* perturbation, and the only known process that is powerful enough to create this is called *inflation*, a hypothetical quantum mechanical process acting 10^{-32} seconds after the Big Bang. If correct, this process is the origin of all structures in the universe today, from galaxies to stars to planets to human beings. This process has not yet been directly measured, but if it is true (and most cosmologists today believe it is), then we are all children of light, created in a storm of quantum perturbations raging 13.8 billion years ago.

To prove the inflationary hypothesis experimentally, we need to detect tensor modes in the CMB polarization field, and this is one of the most active and competitive research fields in contemporary cosmology. Tens of different experiments focus on this single goal, all with the not-so-secret hope of winning a third CMB-based Nobel Prize. Planck's polarization measurements are the best full-sky polarization maps currently available, and a 20° × 20° zoom-in of this map is shown in [03]. For comparison, the second and third panels show perfect scalar (= usual density variations) and tensor (= gravity wave variations) simulations, respectively. Comparing these to the actual observations by eye, it is immediately obvious that the

07   [previous page] Numerical values corresponding to those plotted in [08].
08   Raw voltages as a function of time as measured by three detectors on the QUIET CMB telescope observing from the Atacama desert, Chile, between 2008 and 2010. More than 99% of the observed variations are due to instrumental noise, and only a tiny fraction is due to real structures in the sky.

sky is dominated by the scalar variations. The question, however, is whether there might be a 10%, 1% or 0.1% tensor contribution hidden beneath all those density variations.

In March 2014, a spectacular claim to this effect was announced by an American experiment called BICEP2, deploying the world's most sensitive CMB detectors in a telescope at the South Pole. According to BICEP2's measurements, about 20% of the signal was tensor-like. Unfortunately, when combining the BICEP2 measurements with the Planck observations, it was quickly discovered that the positive measurement was not due to gravity waves from the Big Bang, but something far more prosaic, namely vibrating dust grains in our own galaxy. These also produce polarized light, as shown in [02], with a directional pattern that resembles gravitational tensor modes, as seen in [06]. Although stunningly beautiful, the Milky Way casts long shadows over all experiments searching for the minute inflationary gravity waves. For BICEP2, they proved devastating. The hunt therefore continues at full strength, and with luck we may hope to see the fruits of that work within a decade or two.

As impressive as these hard physical measurements and conclusions are, we are left just as clueless with respect to understanding their deeper origin. In the words of Stephen Hawking, "What breathes fire into our equations?" Why is Nature predictable and well behaved, and why does gravity obey Einstein's field equations? Indeed, what *is* a Law of Nature? These are questions that are just as accessible to philosophers and theologians as to scientists and physicists. But at least one thing is clear to everybody after thousands of years of thinking: Light is the origin of all things.

POSTSCRIPT

After this text was written, the US-led Laser Interferometer Gravitational-Wave Observatory (LIGO) collaboration announced, on 11 February 2016, the world's first direct detection of gravitational waves. Those waves were similar in nature to those discussed above, but were created by two colliding black holes rather than by quantum fluctuations during the birth of the universe. As Galileo opened up a revolutionary new view of the universe based on visible light with the telescope in 1609, LIGO – and hopefully soon several CMB polarization experiments – has opened up an entirely new view based on gravitational waves. We are all indeed very, very lucky to live through such exciting times, being among the very first to see the universe with entirely new eyes!

Steve Northam, International Sales
& Business Development Director, Surrey Nanosystems

VANTABLACK:
WHAT IT IS & WHAT IT ISN'T

ULTIMATE DARK   ABSENCE   NANOTUBES   TOOL   ABSORPTION

Vantablack was created in 2013 by Surrey NanoSystems Ltd, and is now officially recognized as the world's darkest manmade substance, reflecting as little as 0.036% of the light that strikes it.

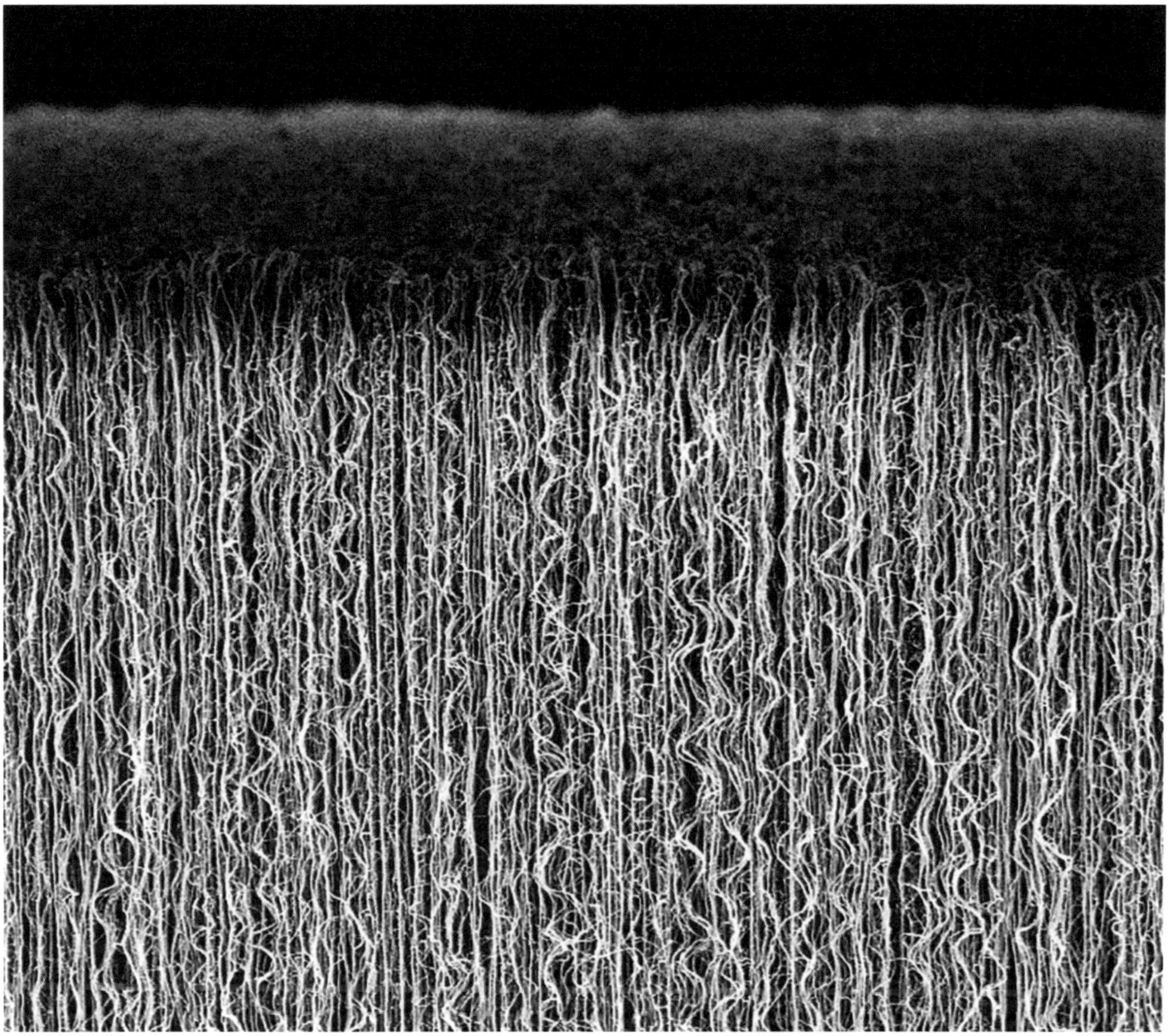

Vantablack is not a black paint, pigment or fabric; but instead is a "carpet" of millions upon millions of incredibly small tubes made of carbon, or carbon nanotubes. Each nanotube in the Vantablack carpet has a diameter of around 20 nanometres (that's about 3500 times smaller than the diametre of the average human hair), and is anywhere from around 20 microns to 200 microns long.

Light energy striking the Vantablack surface enters the space between the nanotubes and is rapidly absorbed as it "bounces" from tube to tube and simply cannot escape, as the tubes are so long in relation to their diameter and the space between them. The near total lack of reflectance creates an almost perfectly black surface [01].

To understand this effect, try to visualize walking through a forest in which the trees are around 3 km tall instead of the usual 10 to 20 metres. It's easy to imagine just how little light, if any, would reach you [02].

01–02

01   A Vantablack surface at × 25 000 magnification. The actual width of material covered within this picture is around 10 microns (10/1000th of a millimetre).
03   [next page] Aluminium foil coated with Vantablack.

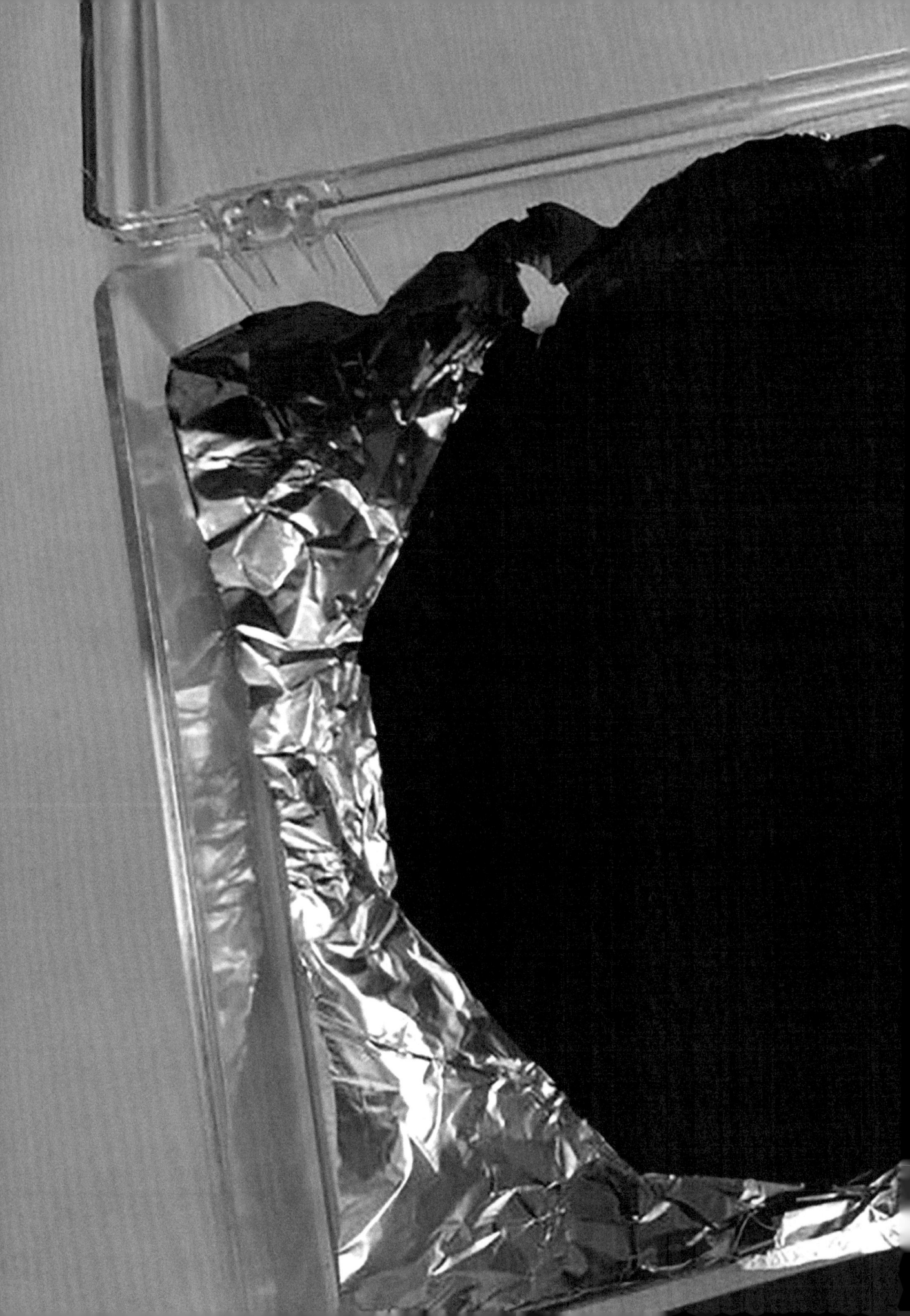

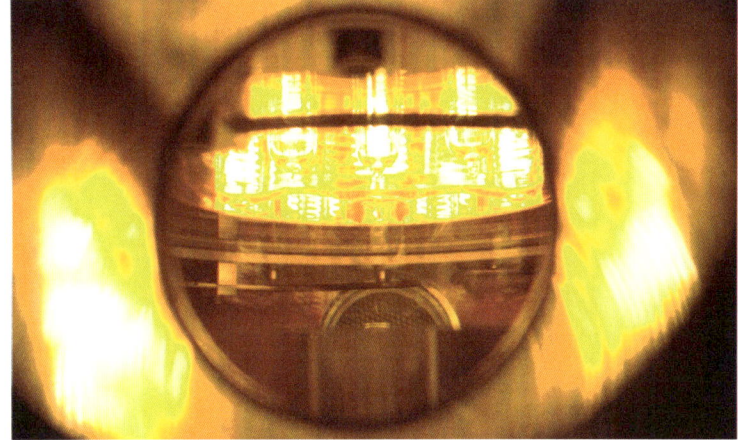

04

The piece of aluminum foil on the previous page [03] provides a good example of the unusual visual effect created by Vantablack. The foil has been creased or wrinkled in a random fashion before being coated with Vantablack on one side. On the uncoated side, the creases are clearly visible, but the coated side appears as a completely flat, black surface (some observers describe it as more like looking into the abyss), as the almost total lack of reflected light from the Vantablack means the human eye is unable to detect that surface. Vantablack is manufactured in a vacuum under an array of powerful lamps that raise the surface temperature to 430°C or higher to allow the nanotube forest to grow [04].

So light itself is used in the production of the world's darkest substance!

François Morellet, artist

DIRECT LIGHT SOURCES IN ART

NEON ARTIFICIAL RHYTHM GEOMETRY

Before the use of electric lighting, the visual arts were designed above all to be looked at in the natural light of day. At the time, artificial lighting seemed inadequate as much for creating as for appreciating works (with the exception, of course, of certain rock paintings).

01

The visual impression of works then was therefore dependent on external light: on its quality, its intensity and its direction. These works existed only by the fraction of light they reflected. The artist was the slave to this unstable element, since one cannot make a work of art with the sun itself.

Soon after the introduction of electric light, it was understood that it could provide a constant, controllable light. This was a first step in the use of electric light in art. In many museums artificial lighting was adopted for paintings and sculptures, setting off the works in a stable way.

The second step was for artists to play with the reflection of this electric light, just as their predecessors had played with the light of day. But this time, they could obtain effects that were much more controllable, more subtle. Moholy-Nagy for example, obtained new effects, which were as much a function of the light source as of the object itself.

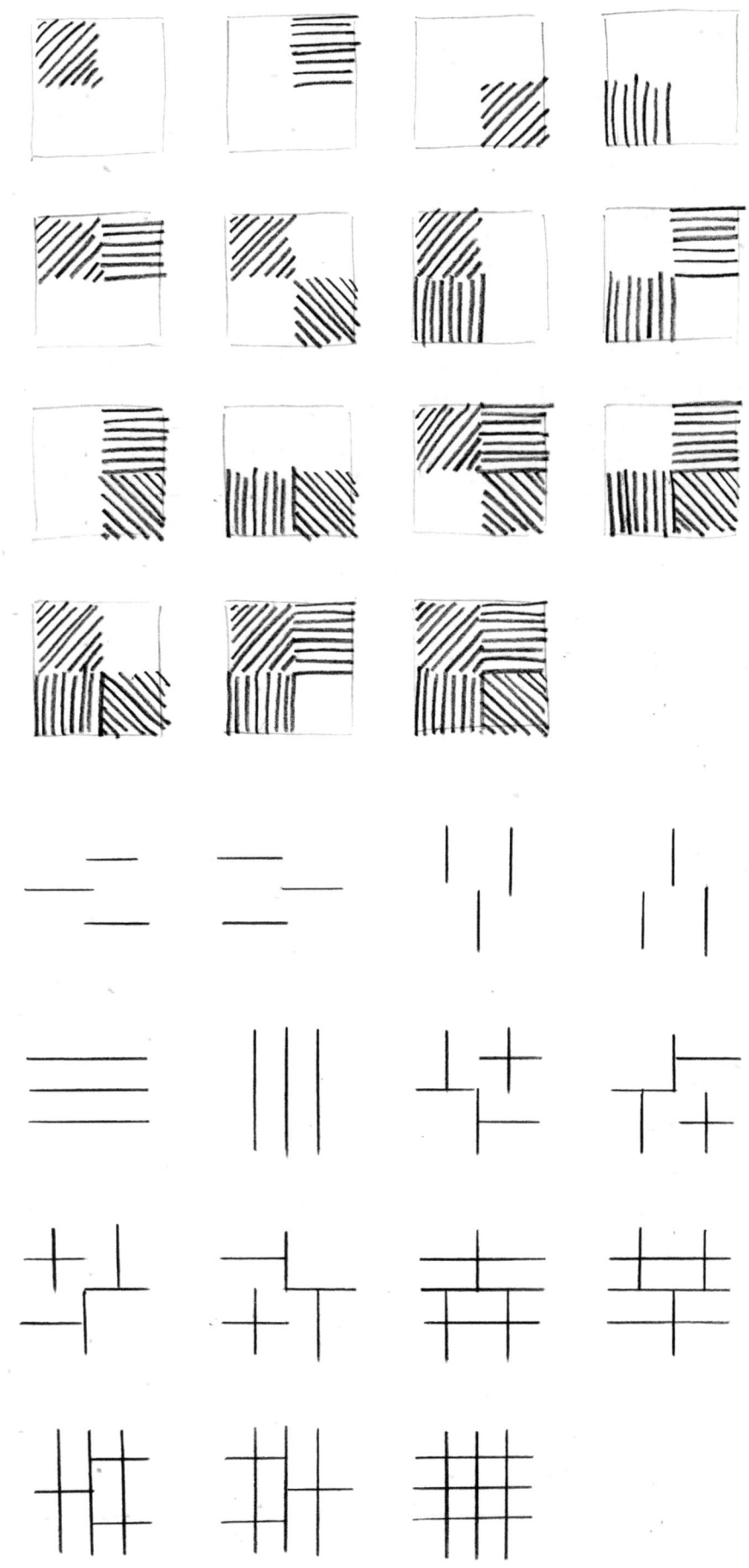

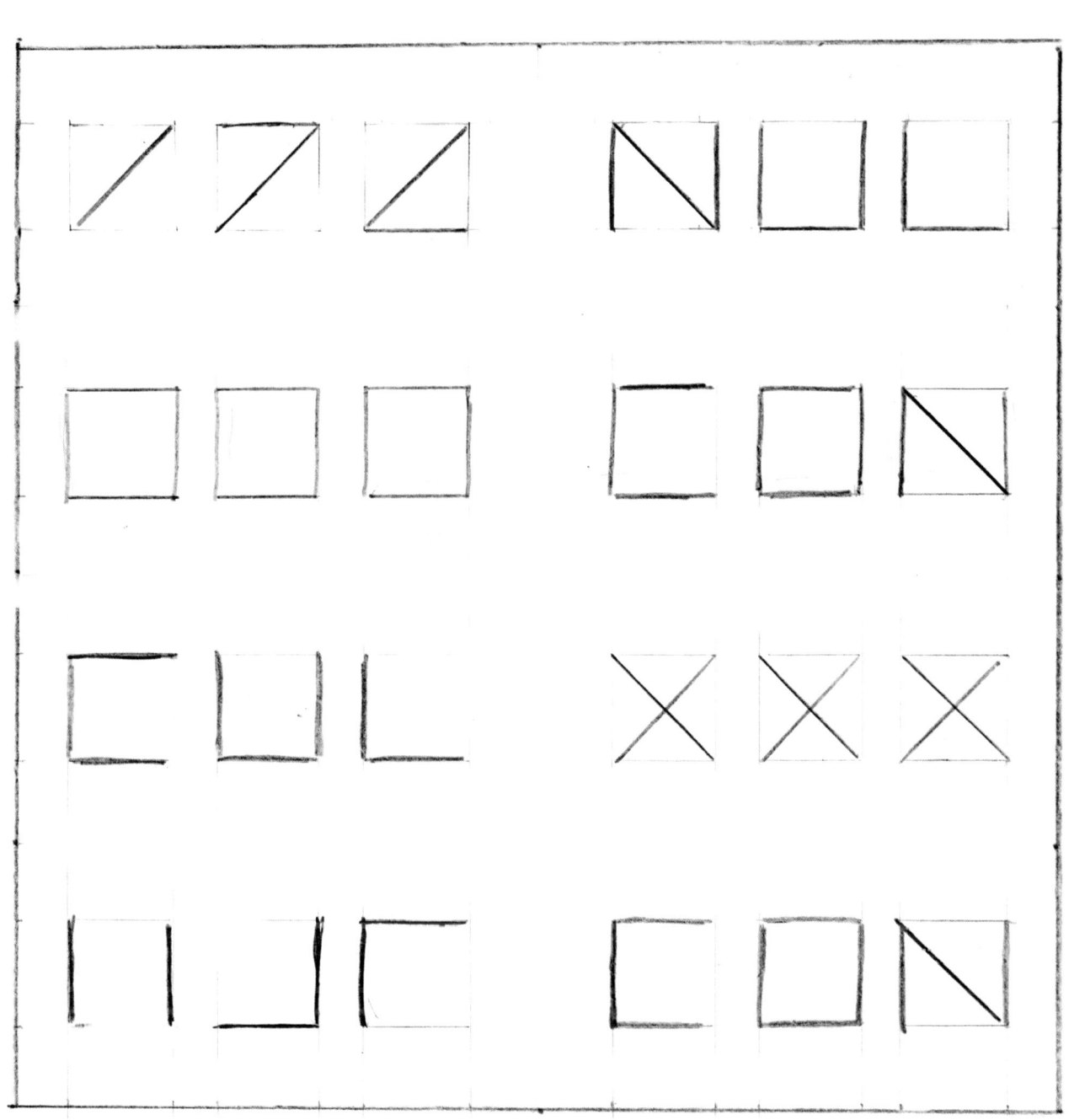

04

01–02 "Néon 0°_45°_90°_135°avec
4 Rythmes Interférents", 1963.
03; 05 "Néon 0°_90°avec 4 Rythmes
Interférents", 1965.
04; 06–08 "Néon avec
Programmation Aléatoire Poétique
Géométrique", 1967.

At the current time, several artists are also using the reflections of artificial light, either on objects or on screens.

We are now coming to third stage, where the light source, rather than its reflection, must be considered as raw material for the artist.

It is only routine and tradition that have hitherto prevented direct sources of artificial light (such as light bulb and neon tubes) from taking the prominent position they deserve in the aesthetic arsenal.

This new material leaves the door wide open to a host of new experiments in visual art: programming, consecutive images, controlled eye movements, luminous rhythms, and whatever else the future holds.

François Morellet, Cholet, France,1966

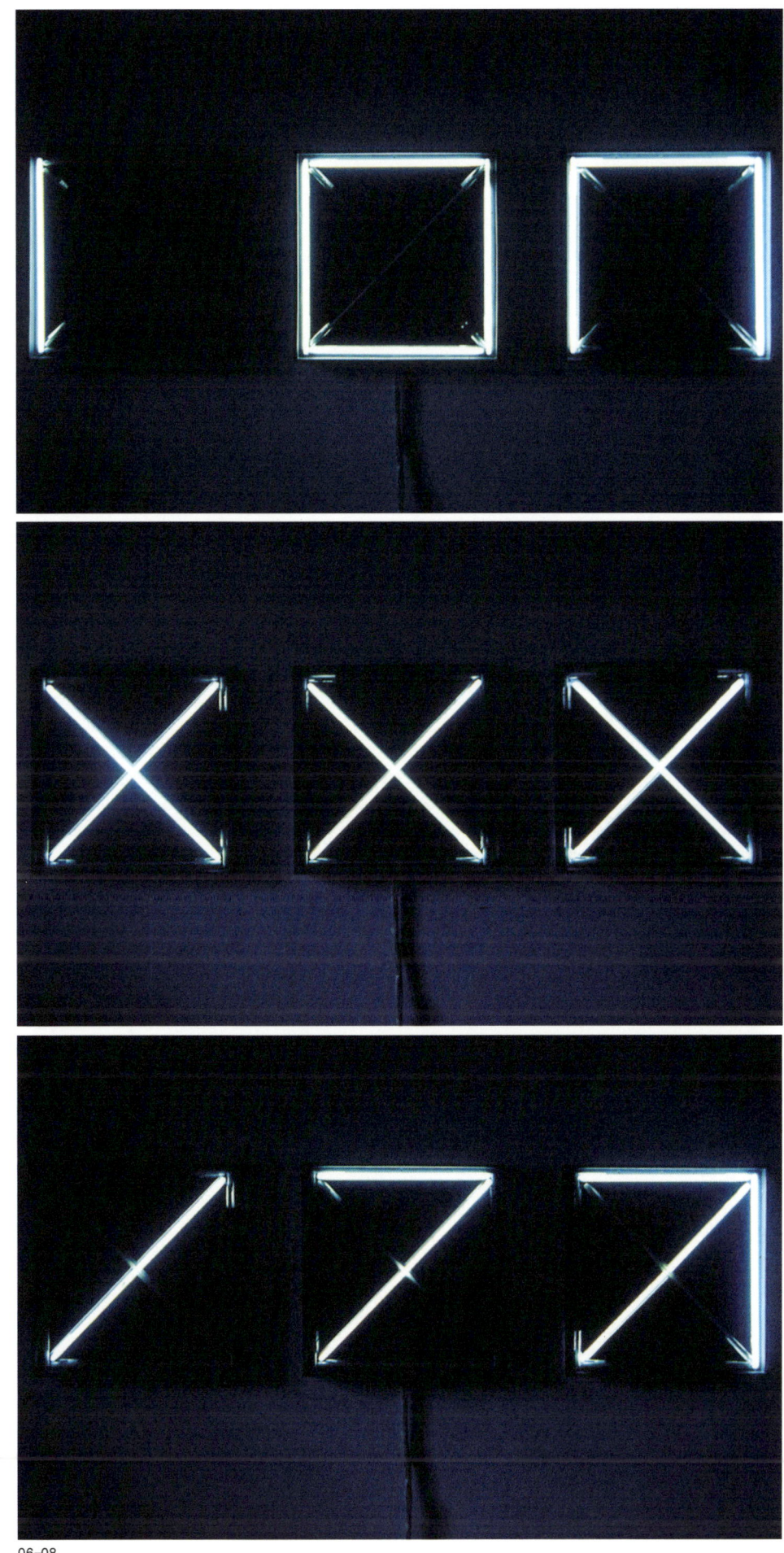

John Day, insect geneticist

GENE DUPLICATION IN BEETLE BIOLUMINESCENCE

GLOW LUCIFERASE LANTERN BEETLES FIREFLIES RAILROAD WORM
EVOLUTION SEDUCTION

In particular species the duplication of genes under copy-number selection is often indicative of species-specific neofunctionalization. At least 152 genes duplicated in the beetle genome *Tribolium castaneum* have single-copy status in all other insects studied. Thus, it has been argued that the expansion of such gene families in beetles, which has frequently been associated with a certain adaptation pressure, may reveal physiologically and phenotypically unique features of beetles. One such unique feature is the phenomenon of bioluminescence that uses a mechanism found in only a small percentage of coleopteran species, and to date, nowhere else in the animal kingdom. Recent evidence has been found to support the hypothesis that bioluminescence, at least in part, has arisen as a result of gene duplication.

BIOLUMINESCENT BEETLE EVOLUTION

Bioluminescence, the production of light by an enzyme-based reaction in a living organism, is evident in a vast range of organisms. These include bacteria, fungi, protists, algae, squid, fishes, as well as terrestrial arthropods. The ability to make light is considered to have evolved many times: based upon a rough estimate of distinct light-producing chemical mechanisms across monophyletic lineages, it has been estimated that bioluminescence has evolved at least 40 times, and likely more than 50 times.

Bioluminescence is typically produced by the oxidation of a light-emitting molecule, generically known as the luciferin, in conjunction with a catalyzing enzyme, either a photoprotein or a luciferase. Although light-emitting reactions are quite distinct in different bioluminescent lineages, all require an oxidation process with molecular oxygen and include the production of light from chemical energy.

Apart from a few species of flies, the great majority of bioluminescent insects are beetles. The oldest primitive protocoleopteran record is from the Early Permian, around 280 MYA (million years ago), but what are regarded as true Coleoptera are not considered to have arisen until the Triassic, around 230 MYA. The vast diversification of the Coleoptera was evident by the Late Jurassic, 155–160 MYA.

Bioluminescence is evident in only a small proportion of beetles: around 3000 of the 370 000 known beetle species are luminescent. All bioluminescent beetles, with the exception of one dubious species, are contained within the Elateriformia. Subdivided into some forty extant families, the series Elateriformia is one of the major groups of beetles within the suborder Polyphaga and contains around 40 000 species.

Bioluminescence in the series Elateriformia is currently classified as occurring in Elateridae and four other families: Omalisidae, Rhagophthalmidae (currently contended to be a part of the Phengodidae), Phengodidae and Lampyridae, which originally all belonged to the superfamily Cantharoidea. All bioluminescent beetles, with the exception of the single staphilinid species, are now currently contained within a single superfamily, the Elateroidea.

The interest in bioluminescence has inspired several studies that investigate the phylogenetic relationships of the Elateriformia, although the relationship between the 18 families of the Elateroidea has yet to be fully resolved as different authors present quite different associations. The only consensus that can be drawn at present is that bioluminescent beetle families appear not to have descended from a common ancestor and upon first inspection appear, in some cases, to have arisen independently.

BEETLE BIOLUMINESCENCE

The most studied bioluminescent system is catalyzed by firefly luciferase, in particular the North American *Photinus pyralis* firefly luciferase (Luc) (monooxygenase, [EC 1.13.12.7]). A two-step scheme was proposed for the overall reaction of firefly bioluminescence.

1 $E + LH2 + ATP + Mg^{2+} \rightarrow$
   $E{\cdot}LH2{-}AMP + PPi + Mg^{2+}$

2 $E{\cdot}LH2{-}AMP + O_2 \rightarrow$
   $E{\cdot}L + CO_2 + AMP + Light$

Both steps are catalyzed by the enzyme luciferase (E). In the first stage luciferin (LH2) is converted into a luciferyl adenylate (LH2-AMP) by ATP in the presence of Mg^{2+}. In the second step, luciferyl adenylate is oxidized by molecular oxygen, resulting in the emission of light and the production of oxyluciferin (L).

The luciferase was first purified from the firefly *Photinus pyralis* in 1956 and later cloned and expressed in 1985, providing the basis for mass production of luciferase *in vitro* and the further characterization of the enzyme through mutagenesis studies in the coming years. To date the luciferase cDNA has been characterized from over 20 bioluminescent beetle taxa, and extensive information has been collated about these enzymes.

ADENYLATE-FORMING PROTEIN FAMILY

A number of amino acid residues in beetle luciferase have been found to be highly conserved in a range of related enzymes which are classified as belonging to a large superfamily of adenylate-forming enzymes (PFAM00501). The adenylate-forming proteins catalyze a two-step reaction converting an organic acid to a CoA thioester. This mode of substrate activation is commonly used by adenylate-forming enzymes such as acyl-CoA ligases, acetyl-CoA synthetases, non-ribosomal peptide synthetases (NRPSs) and aminoacyl-tRNA synthetases, as well as luciferase. These enzymes are relatively large, ranging in size from 500 to 700 residues. Structurally they are composed of two domains, an N-terminal domain of 400–550 residues and a smaller C-terminal domain of 100–140 residues. An active site is situated at their interface. Members share limited sequence homology of 20–30%, however, several well-conserved sequence motifs have been identified between members, and three principal motifs have been attributed with an adenylation function. Of particular note is the invariant residue K^{529}, which appears to be important in the adenylation step.

These enzymes activate a variety of different substrates, including aromatic acids, acetic acid and long-chain fatty acids, to the corresponding enzyme-bound acyl-adenylates, which are then transferred to the thiol group of CoA. The two half-reactions occur in a ping-pong mechanism. A domain alternation mechanism has been proposed for these enzymes. Upon completion of the initial adenylation reaction, the C-terminal domain of these enzymes undergoes a 140° rotation to perform the second thioester-forming half-reaction.

It has been speculated that beetle luciferase may have evolved from an ancestral fatty acyl-CoA synthetase as firefly luciferase retains this activity *in vitro*.[1, 2] As such, beetle luciferin may not itself have originally been the substrate for the ancestral luciferase, but rather a "luciferin-like" molecule, with beetle luciferin appearing as a substrate later on in the evolution of beetles. In support of this, dehydroluciferin, differing from luciferin by only two hydrogen atoms and inactive for chemiluminescence, can be efficiently ligated to CoA by firefly luciferase.[3] Luciferase may still function as a fatty acyl-CoA synthetase involved in the oxidation of fatty acids in the peroxisome of beetles. Interestingly, it was shown that firefly luciferase had a marked preference for fatty acids such as arachidonic acid.[1] This may be unsurprising as arachidonic acid, although typically occurring in very small amounts in the phospholipids of terrestrial insects, has been found in very high levels in the tissue lipids of adult fireflies.[4]

LUCIFERASE GENE DUPLICATION IN FIREFLIES AND GLOWWORMS

Paralogous luciferase-like sequences have been identified from the Japanese firefly *L. cruciata*,[2] suggesting gene duplication of luciferase-like sequences in bioluminescent beetle genomes. Despite extensive sequence identity of the *L. cruciata* luciferase-like genes to the *bona fide* luciferase, the two paralogous enzymes revealed no bioluminescence activity. Furthermore, only one gene product exhibited long-chain fatty acyl-CoA synthetic activity. More recently Oba et al found that the yellow light emitted from the adult lanterns is a product of a different enzyme to that producing green light in the eggs and pupae. They identified a luciferase isotope from *Luciola cruciata* that was functional and expressed in firefly eggs (Oba et al, 2010).

Similarly the same luciferase isotope was discovered from *Luciola lateralis* (Oba et al, 2013). This second luciferase, denoted as *Luc2*, was expressed in the eggs and pupae, which glowed dimly at a wavelength of 539 nm. The luciferase found in the adult lantern, *Luc1*, produced a different wavelength, that of 550 nm.

It was subsequently proposed that luciferase has arisen from a gene duplication event of an ancestral fatty acyl-CoA synthetase and functionally diverged to acquire a novel bioluminescent function.[2] Luciferase orthologues found in the nonbioluminescent mealworm *Tenebrio molitor* exhibited no bioluminescent activity and were reported to have acyl-CoA synthetase activity.[5] However, a luciferase-like sequence identified in the nonbioluminescent mealworm

Zophobas morio was found to exhibit weak biolu-minescent activity (Viviani et al, 2009).

Crowson proposed that luminescence must have originated as an accidental by-product of a chemical reaction serving an alternative purpose and that non-adaptive luminescence is unlikely to persist for long in the evolutionary timescale (Crowson, 1981). It seems feasible that luciferase, the only oxygenase in the adenylate-forming enzyme superfamily, may have originally played a role in controlling levels of oxidative stress with or without a luciferin. The first light generated from the primeval bioluminescent beetles would, as Crowson suggests, need to be adapted for a role that would ensure its survival as a trait over time.

Duplications in particular gene families are often regarded as an important source of evolutionary novelties that contribute to inno-vative phenotypic traits and biological func-tions specific to certain groups of organisms.[6] Currently, the divergence of two paralogues after duplication is considered to follow one of three routes. The most likely outcome of a duplication event is nonfunctionalization when one copy first becomes a pseudogene and eventually becomes extinct,[7] whereas the second copy retains the original function. The other, less frequent but nonetheless essential evolutionary scenarios are neofunctionalization and subfunctionalization. In the event of neofunctionalization, one paralogue retains the original function, whereas the other evolves a new function during a period of rapid, nearly neutral evolution. Under the subfunc-tionalization model multiple functions of the ancestral gene are separated between paralogues, both of which evolve under purifying selection.

Based upon the extensive knowledge of beetle luciferase, these enzymes and their lucif-erase-like orthologues present a rare opportu-nity to investigate the evolution of functional diversification in an insect gene family.

A number of phylogenies reconstructed from beetle luciferase sequences are present in the literature. Despite the reports of two or more luciferases in *Photuris,* many authors have chosen to ignore this situation and only seen fit to select one, *P. pennsylvanica,* luciferase paralogue for their phylogenies. It has become evident that a single gene luciferase phylogeny for the Lampyridae is redundant and a complex evolutionary pattern exists of luciferase and luciferase-like paralogues in lampyrid genomes (Day et al, 2009) [01].

Good phylogenetic support has been found for two major clades in beetle adenylate-forming enzyme evolution. The divergence of two paral-ogues after duplication can follow one of three

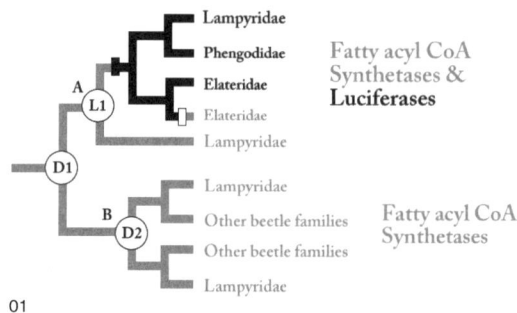

01

routes, nonfunctionalization, neofunctionalization and subfunctionalization. The most expected outcome for a duplicated gene is nonfunctionali-zation. However, with regards to bioluminescence in beetles the most parsimonious scenario seems to be a single neofunctionalization event in the ancestral Elateriformia, followed by subfunction-alization and/or further duplication events [01].

An early duplication event denoted by D1, possibly after the Elateriformia formed, has resulted in a neofunctionalization event generating a dual-functioning luciferase in all the bioluminescent beetle families. However, luminescent activity has been lost in nonlumi-nescent elaterid species. A later duplication in clade A, denoted by L1, is currently only evident in the Lampyridae.

Nonfunctionalization may have occurred in certain beetle lineages, such as the Canthar-idae, which are closely related to the Lampyridae, where studies show no evidence for a clade A paralogue. Thus the ancestral clade A paral-ogue in nonbioluminescent beetles may have been become silenced by degenerative muta-tions, as is the case with most duplicated genes. However, individual paralogues of both A and B clades have proven functionality and exhibit either a bioluminescent activity or an acyl-CoA synthetase activity (Oba et al, 2006b). Thus the early duplication event resulting in both clades would have given rise to functional paralogues which would have been fixed and subsequently evolved by genetic drift and positive selection. In particular, positive selection is thought to drive the fixation of a duplicate gene that has gained a new function through acquisition of a beneficial mutation, this is the process referred to as neofunctionalization.

Oba et al, (2006b) presented a putative schematic for the evolutionary descent of firefly luciferase from an ancestral fatty acyl-CoA synthetase. They speculated a single gene dupli-cation event from an ancestral gene resulting in three paralogues, one of which subsequently evolved a new function of bioluminescence. This scenario is classified as the luciferase neofunc-tionalization scenario.

RAILROAD WORM AND CLICK BEETLE LUCIFERASES

The bioluminescent mechanism in the railroad worms (Phengodidae) and luminescent click beetles (Elateridae) is considered to be the same as that found in fireflies (Lampyridae). Each mechanism is dependent upon ATP, luciferin, Mg^{2+} and the enzyme luciferase to create light. Beetle luciferin is regarded to be the same structure in Lampyridae, Phengodidae, and luminescent elaterids.[8, 9] Despite these similarities the difference in colours of light produced in these families is quite dramatic.

In lampyrids the light is limited in a range from green to yellow (l_{max} 538–584 nm). However, bioluminescent click beetles have three light organs; a pair of dorsal, oval light organs on the pronotum which emit a green light (l_{max} 536–559 nm) and a ventral organ located on the first abdominal segment which ranges in colour from green through to orange (l_{max} 549–594 nm). In railroad worms the number of lanterns increases, with eleven pairs of luminous organs located dorso-laterally along the abdominal and thoracic segments. These emit green through to orange light (l_{max} 535–592 nm) and are present in both adults and larvae. In addition, some species, such as the railroad worm *Phrixothrix*, have a luminous organ on the head which emits red light (l_{max} 600–620 nm). These colour differences are a result of amino acid differences in the luciferase protein.

In 1998 the luciferase from *Rhagophthalmus ohbai* was characterized.[10] Although *R. ohbai* is currently classified in its own family, the Rhagophthalmidae, opinion is still divided as to its placement. In the past it has been contained in the Phengodidae and the Lampyridae. The *R. ohbai* luciferase shares greatest sequence identity with the Phengodidae luciferase sequences but this comparison is currently limited to one species.

EVOLUTION OF CLICK BEETLE BIOLUMINESCENCE

Around 9000 described species of the Elateridae are widely distributed across the world. However, unlike the families Lampyridae and Phengodidae, only a small proportion are bioluminescent. Just 200 elaterid species, occurring within the Pyrophorini and Campyloxenini tribes of the subfamily Agrypinae, are considered to be luminescent.

Elaterids are commonly known as click beetles. When an elaterid is resting on its back, a hinge mechanism is locked to store elastic energy in the body. Upon release the beetle is launched into the air, often accompanied by an audible clicking sound.

The adult bioluminescent click beetle usually have two sets of lanterns; a pair of dorsal thoracic lanterns which emit yellow to green light, and a third ventral abdominal lantern emitting orange to green light.

It has been hypothesized that in click beetles abdominal lantern bioluminescence is an optical signal for sexual communication, with males switching on the lantern in flight whilst looking for a mate. Light from the dorsal lanterns, however, warns potential predators and may even provide illumination in flight.

As larvae, luminescent elaterids usually emit green light from prothoracic lanterns but also from several small spots along the body. Unlike the Lampyridae and Phengodidae, not all members of the family Elateridae exhibit bioluminescence. Furthermore, these luminous click beetles are only found in tropical and subtropical America and Melanesia.[11]

The luciferin in luminous elaterids has been shown to be identical to firefly luciferin but also to be in higher concentrations than in some fireflies,[12] which may account for previous records of the click beetle genus *Pyrophorus* having a great luminosity than lampyrids.[13]

However, no luciferin was detected in four non-luminous click beetle species,[12] one of which, *Agrypnus binodulus*, also expressed a luciferase-like protein which exhibited fatty acyl-CoA synthetase activity but showed no luciferase activity.[14] Later, Oba et al, (2009) established that replacing a single amino acid with serine restored luminescent activity of the enzyme. This is indicative of the loss of the luciferin biosynthetic pathway and luciferase activity in nonluminescent elaterids.

Shortly after the publication of the first firefly luciferase sequence, four different luciferase sequences were characterized from a single click beetle species, *Pyrophorus plagiophthalamus*.[15] Sixty beetles were used to construct a cDNA library from which the luciferases where characterized. It was not evident at the time whether these different enzymes could be found in a single beetle or whether the dorsal and ventral lanterns were under different genetic control. One luciferase clone was characterized from *Pyrearinus termitilluminans,* which produced a blue-shifted bioluminescence. Although one clone was evaluated, four other clones were bioluminescent but unfortunately were not characterized. Additional luciferase genes may exist in the genome of *P. termitilluminans* which are yet to be identified.

01   Schematic representation of relationships between paralogous adenylated-forming enzymes in beetle genomes. Four groups of paralogues have been identified in lampyrid genomes (Day et al , 2009). Ancestral ancient beetle duplication events prior to the diversification of the Elateriformia are highlighted at points D1 & D2. A Lampyridae-specific duplication event is denoted at point L1. A point of origin of bioluminescence is predicted to have occurred prior to the diversification of the Lampyridae and Phengodidae and possibly even the Elateridae, and is denoted by a solid black vertical bar. A loss of bioluminescent function is denoted within the Elateridae by a white vertical bar.

In 2003 Stolz et al conducted a large study on the same species in Jamaica and found, by comparing genomic clones with cDNA sequences, there were two different genes controlling bioluminescence independently in the dorsal and ventral lanterns.[16] Stolz et al found that the luciferase sequence data seemed to imply an exchange event from the dorsal to the ventral luciferase locus on Jamaica. They extrapolated from this that the ancestral bioluminescent colour state in *P. plagiophthalamus* for the ventral organ was green.[16] This exchange event was followed by a series of substitutions in the ventral luciferase locus of *P. plagiophthalamus* that selectively shifted the colour of the ventral organ from green toward longer wavelengths, producing a recently derived ventral orange allele arising on the eastern side of the island.[16]

This intergenic exchange was later examined in other bioluminescent click beetles and found to be a general phenomenon in *Pyrophorus* species.[17]

CONCLUSIONS

Strong evidence supports gene duplication of an ancestral fatty acyl-CoA synthetase during beetle evolution, followed by a neofunctionalization of this gene or genes to a dual-functioning enzyme capable of creating light emission by the oxidation of firefly luciferin. In many firefly species multiple duplication events are evident. In click beetles gene duplication has resulted in a phenotypic shift in colour emission and subsequent differential expression in dorsal and ventral lanterns. The extent of gene duplication and its impact on bioluminescence in beetles is still to be fully resolved, but to date a solid body of research presents an excellent foundation for future work.

1   Y. Oba, M. Ojika, S. Inouye (2003), "Firefly luciferase is a bifunctional enzyme: ATP-dependent monooxygenase and a long chain fatty acyl-CoA synthetase", *FEBS Lett*, 540:251-254

2   Y. Oba, M. Sato, Y. Ohta, S. Inouye (2006), "Identification of paralogous genes of firefly luciferase in the Japanese firefly, Luciola cruciata", *Gene*, 368:53-60

3   R. Fontes, A. Dukhovich, A. Sillero, M.A.G. Sillero (1997), "Synthesis of dehydroluciferin by firefly luciferase: effect of dehydroluciferin, coenzyme A and nucleoside triphosphates on the luminescent reaction", *Biochemical and Biophysical Research Communications*, 237:445-450

4   A.R. Nor Aliza, J.C. Bedick, R.L. Rana, H. Tunaz, W.W. Hoback, D.W. Stanley (2001), "Arachidonic and eicosapentaenoic acids in tissues of the firefly, *Photinus pyralis* (Insecta: Coleoptera)", *Comp Biochem Physiol A Mol Integr Physiol*, 128:251-257

5   Y. Oba, M. Sato, S. Inouye (2006), "Cloning and characterization of the homologous genes of firefly luciferase in the mealworm beetle, *Tenebrio molitor*", *Insect Mol Biol*, 15:293-299

6   S. Ohno (1970), *Evolution by gene duplication*, Springer, New York

7   M. Nei, A.K. Roychoudhury (1973), "Probability of fixation of nonfunctional genes at duplicate loci", *Am Nat*, 107:590-605

8   H.H. Seliger, W.D. McElroy (1964), "The colours of firefly bioluminescence: enzyme configuration and species specificity", *Proc Natl Acad Sci USA*, 52:75-81

9   V.R. Viviani, E.J.H. Becham (1993), "Biophysical and biochemical aspects of phengodid (railroad-worm) bioluminescence", *Photochem Photobiol*, 58:615-622

10   M. Sumiya, V.R. Viviani, N. Ohba, Y. Ohmiya (1998), "Cloning and expression of a luciferase from the Japanese luminous beetle *Ragophthalmus ohbai*", in A. Roda, M. Pazzagli, L.J. Kricka, P.E. Stanley (eds), *Bioluminescence and Chemiluminescence: Proceedings of 10th International Symposium*, Bologna, Italy, 433-436

11   J.N.L. Stibick (1979), "Classification of the Elateridae (Coleoptera): relationships and classification of the subfamilies and tribes", *Pacific Insects*, 20:145-186

12   Y. Oba, T. Shintani, T. Nakamura, M. Ojika, S. Inouye (2008), "Determination of the luciferin contents in luminous and non-luminous beetles", *Biosci Biotechnol Biochem*, 72:1384-1387

13   H.C. Levy (1998), "Greatest Bioluminescence", in J. Walker (ed) *University of Florida Book of Insect Records*, University of Florida, Florida, USA, 72-73

14   Y. Oba, K. Iida, M. Ojika, Inouye S (2008), "Orthologous gene of beetle luciferase in non-luminous click beetle, *Agrypnus binodulus* (Elateridae), encodes a fatty acyl-CoA synthetase", *Gene*, 407:169-175

15   K.V. Wood, Y.A. Lam, H.H. Seliger, W.D. McElroy (1989), Complementary DNA coding click beetle luciferases can elicit bioluminescence of different colours", *Science*, 244:700-702

16   U. Stolz, S. Velez, K.V. Wood, M. Wood, J.L. Feder (2003), "Darwinian natural selection for orange bioluminescent colour in a Jamaican click beetle", *Proc Natl Acad Sci USA*, 100:14 955-14 959

17   Feder J, L. , Velez S (2009), "Intergenic exchange, geographic isolation, and the evolution of bioluminescent colour for *Pyrophorus* click beetles", *Evolution*, 63:1203-1216

Roos van Haaften, artist

AIN'T WHAT THE MOON DID

GREY CLARITY BEAUTY GEOMETRY SPACE SHADE

In Western living spaces we seem to have a penchant for light and clarity. Light is associated with progress, safety and beauty. The more light, the better. In architecture we desire high ceilings and glass surfaces. Skyscrapers reach for the sun and parlours have wide views of the garden. Everyone seems to enjoy the sun beaming through a window. Our sense of beauty and even our mental state depend on light and brightness.

This is a predominantly Western approach to aesthetics, as stated by Jun'ichirō Tanizaki in his essay "In Praise of Shadows" (1933). Tanizaki explains passionately how Asian cultures appreciate shadow and dimness over clarity and visibility: "We find beauty not in the thing itself, but in the patterns of shadows, the light and the darkness, that one thing against another creates."[1]

Through architecture, food, drama and garden design, Tanizaki leads us into the subtle nuances of grey. Beauty is, according to Tanizaki, not easily seen in bright light. He mentions the aesthetics of an old, wrinkling face, the dimmed reflections in tin and unpolished gold, and the importance of dark spaces within the bright green of the stacked vegetation in an Oriental garden. Tanizaki connects the phenomena of shadow with feeling at home, wisdom and security. "In making ourselves a place to live, we first spread a parasol to throw a shadow on the earth, and in the pale light of the shadow we put together a house."[2] The essay can even be read as a lamentation about the upcoming use of electric light. But moreover, Tanizaki points out how shadow and darkness are generally considered, as physicists tend to, as an *absence* of light, when in fact he (and I as an artist) consider shadows and darkness as *tones* of light.

Embracing the Western as well as the Oriental approach to light and space generates an abundance of tones to work with; from bright white to deep black – and all the shades that lie in between.

1   Jun'ichirō Tanizaki (1977), *In Praise of Shadows*, 30, Leete's Island Books, Stony Creek
2   idem 17

01–08   [next pages] Profile spots on white background, ink drawings, lenses.

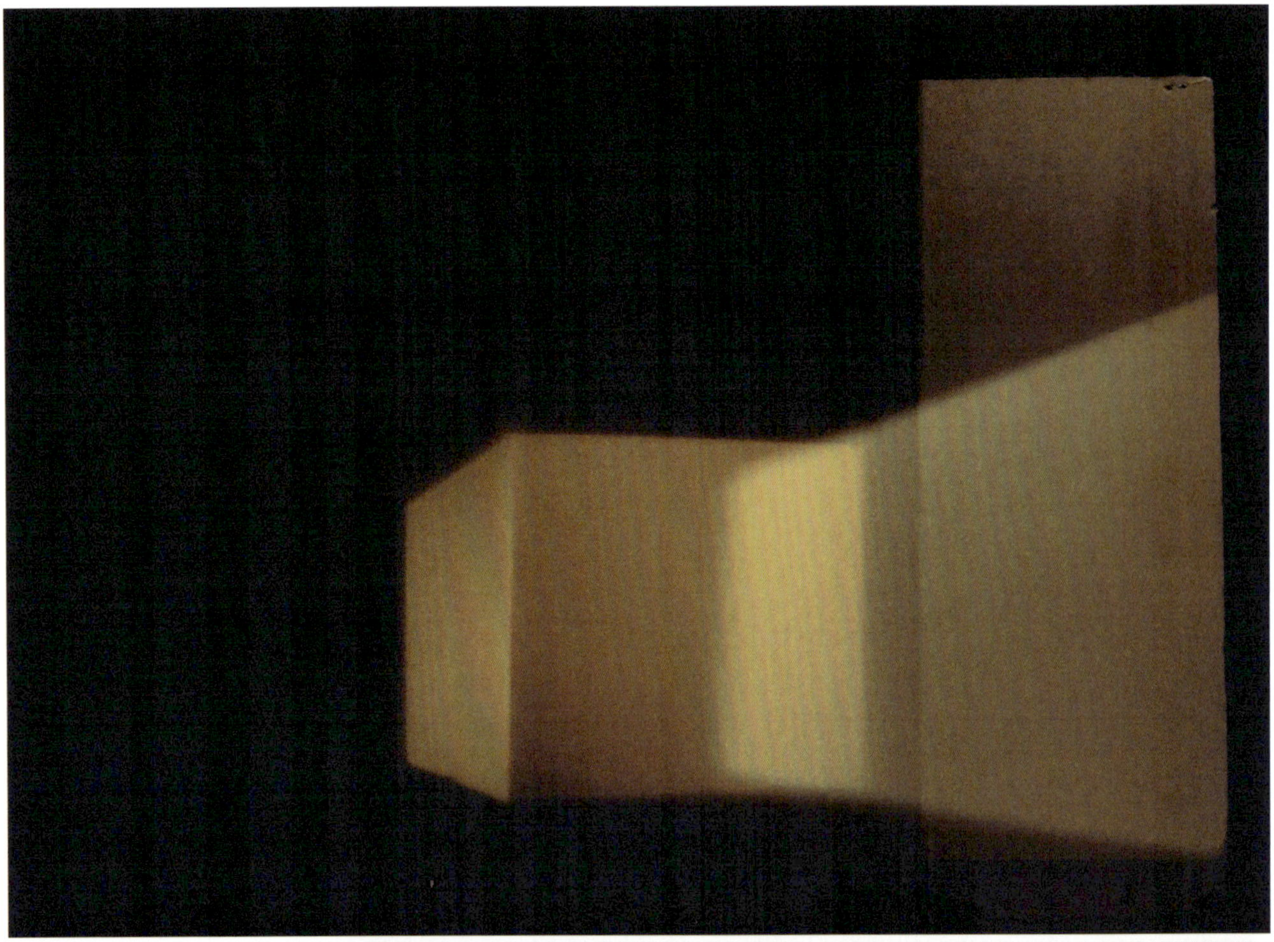

01–02

03–04

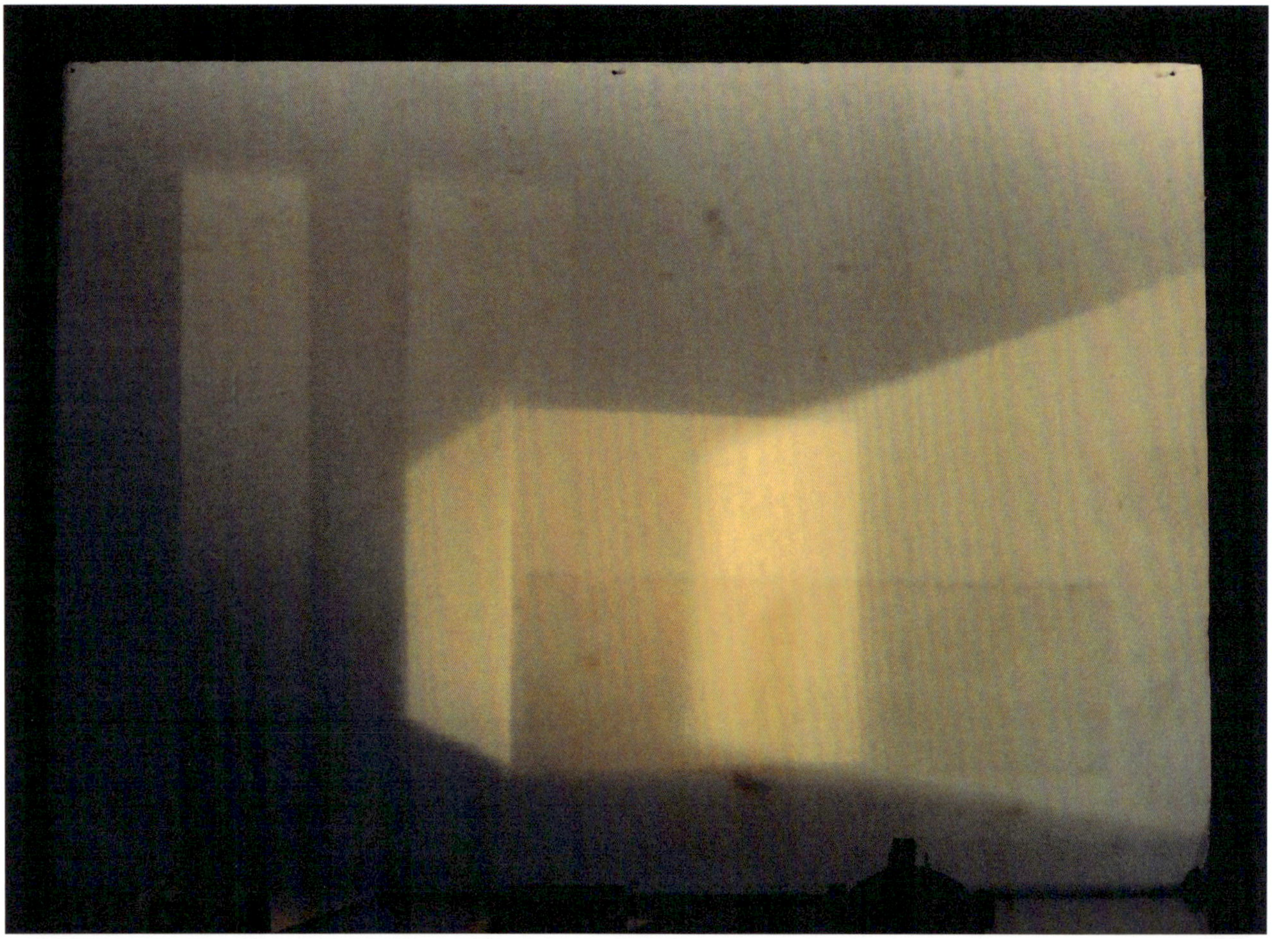

05–06

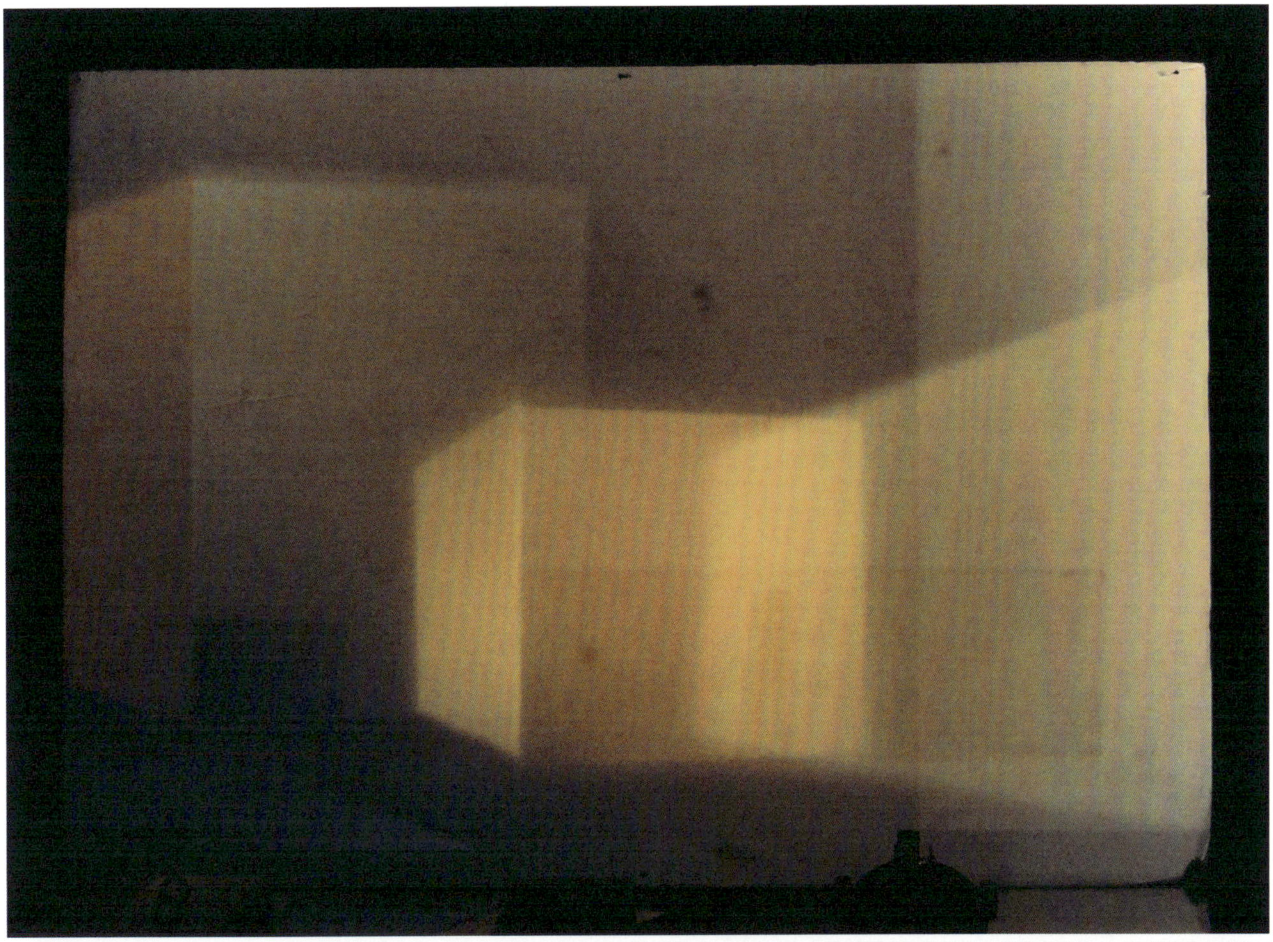

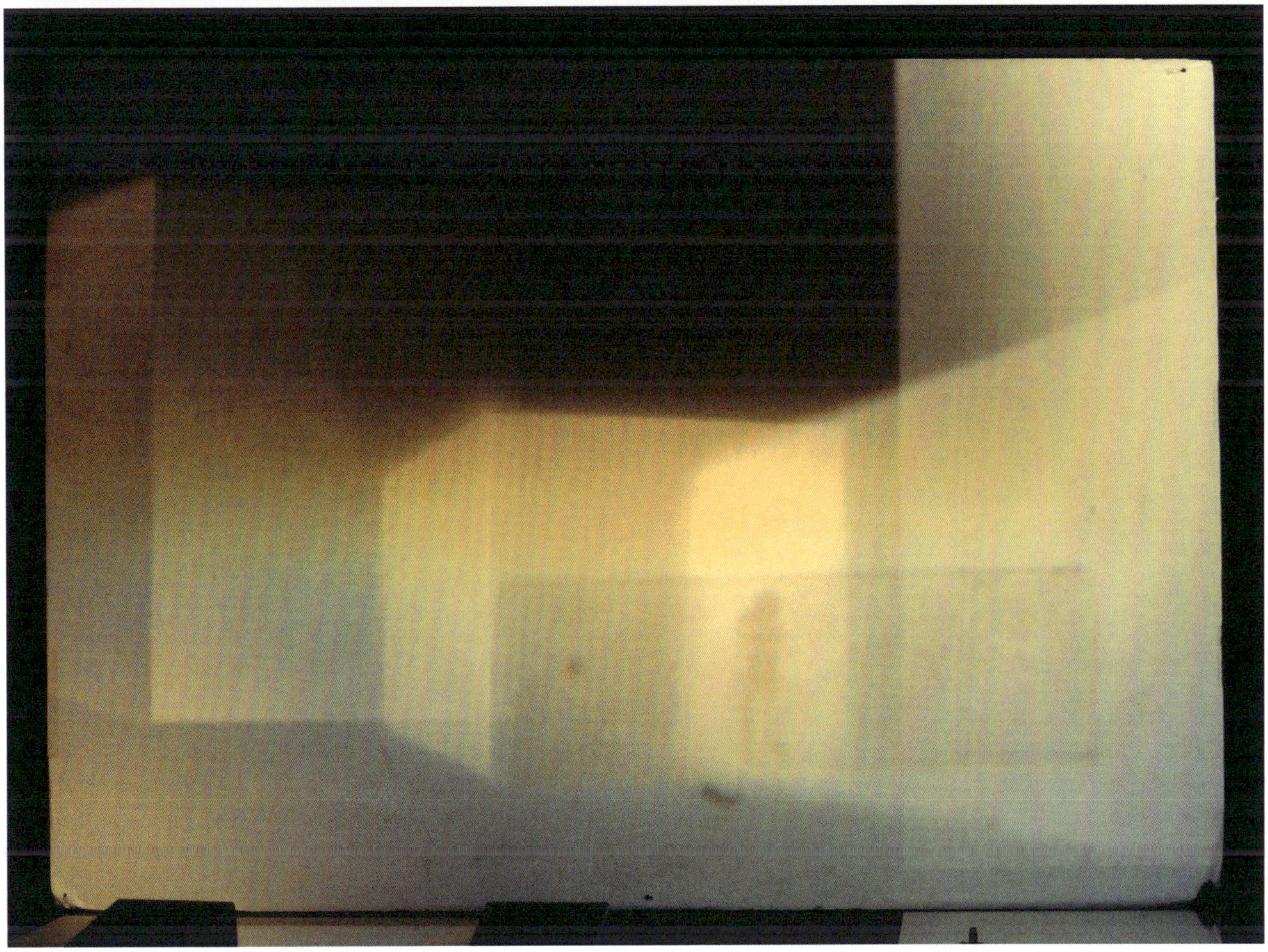

07–08

Bianca van der Zande, topic owner, Human Centric Lighting at Philips Research

LIGHTING THE WAY TO AN AGE-FRIENDLY WORK ENVIRONMENT

MOOD   NEEDS   PERFORMANCE   SPECTRUM   CUSTOMIZED   WORKPLACE

With the changing nature of work and an ageing workforce, employers have to focus on workplace design. That includes lighting.

It's a fact. Organizations have to do more with less. They want to save money, trim staffing levels, cut back on office space and reduce their carbon footprint. At the same time, they want to increase productivity and boost people's wellbeing.

Even as they try to get more out of a reduced workforce, businesses are grappling with the changing nature of work. We spend more time indoors than ever before. With our dependence on PCs and mobile devices these days, we do more and more of our work on screens. We also have conflicting needs – for open workspaces that promote collaboration on the one hand and, on the other hand, private spaces where we can concentrate. A study in 2008 found that workplaces that sacrifice individual focus in favour of collaboration result in decreased effectiveness of both.[1]

And while all this is happening, organizations are facing one of the biggest challenges of our time: we're all living and working longer. Today 30–50% of office employees are older than 45.[2]

These changes mean that employers have to focus on the ergonomic design of our workplaces if they want to maximize their human resources. Employee salaries represent most of an organization's costs. For example, they can amount to 84% of the cost per square foot of a commercial building,[3] or 70 times the amount spent on energy.[4] So it's reasonable to assume that a small increase in productivity is likely to offset the costs of a better working environment. Conditions at work affect our physical and mental wellbeing. These in turn are believed to have an impact on our work capacity, productivity, sick leave, job satisfaction and company loyalty.

But what makes for an optimized work environment? Human-centric design goes beyond ergonomic furniture and equipment, comfortable room temperatures and acceptable noise levels. The lighting in the work environment is important as well. Natural daylight, illumination of the workspace, reduced lighting contrasts, lack of visual clutter and use of colours, accents and patterns all have an enormous effect on how we think and feel at work.

THE IMPORTANCE OF GOOD OFFICE LIGHTING

In functional terms, light enables us to see both small details and the wider world around us. The most obvious effect of light on humans is to enable vision and the performance of visual tasks. For more than 150 years, scientists considered rods and cones to be the only photo-receptor cells in the eye. These photoreceptor cells regulate the visual effects. When light reaches these cells, a complex chemical reaction occurs. The chemical that is formed creates electrical impulses in the nerve that connects the photoreceptor cells to the visual cortex at the back of the brain. In the visual cortex the electrical impulses are interpreted as "vision". The rods operate in low-level light situations and do not permit colour vision. The cone system, functioning under normal daytime lighting conditions, is responsible for sharpness, detail and colour vision. The sensitivity of the cone and rod systems varies with varying wavelengths of light, and thus with varying colours of light.

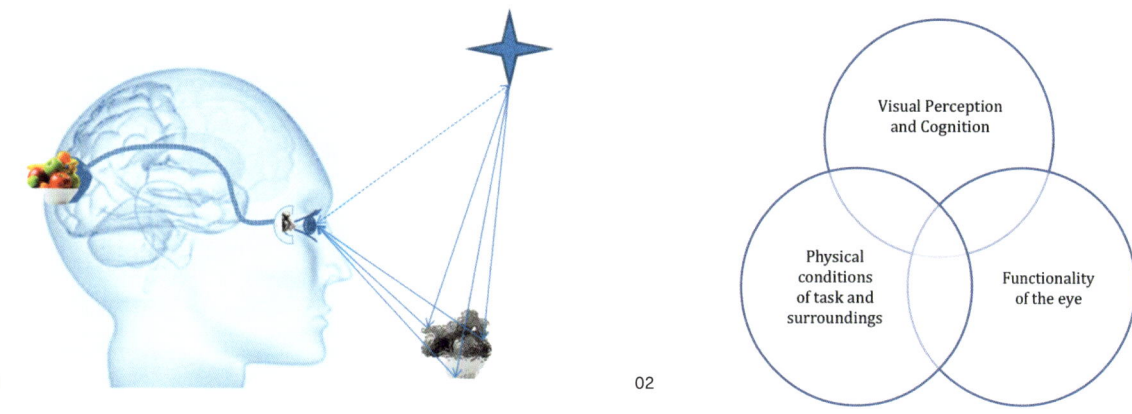

01

02

Studies[5] show that the nature of the task – as well as the amount, spectrum and distribution of the light – determines the level of visual performance that is achieved. The current European standard for writing, typing, reading and data processing in offices is 500 lux. This is only the minimum though, and may not be enough for more intense or sustained visual work, or ageing workers. One laboratory study demonstrated that 1000 lux supports highly concentrated work while dimmer lighting enhances creative thinking.[6]

Inadequate lighting can lead to visual discomfort.[7] What's more, because the eye muscles are connected to the neck and back muscles, bad lighting can result in neck pain, headaches and fatigue as the brain struggles to interpret visual information.[8]

Inadequate lighting also has strong implications for the ageing workforce. We all experience changes to our eyes as we get older. The lens loses flexibility and becomes cloudier, making images more blurred. Ageing also affects the vision in terms of glare and when moving from dark to light areas.[9] As a result, a 60-year-old person needs significantly more light than a 20-year-old to see the same visual detail (see [03]) and

01–02 The ability to see small details is a result of the complex system of light reflected from the task that is captured by the eye and sent to our brain. What you see depends on the functionality of the eye, visual perception and cognition, as well as the physical conditions of tasks and surroundings.

will also have more problems seeing small details on the computer screen, as visualized in [04]. As compared with a 20-year-old the visual acuity at a distance of 60 cm drops by 80%.[10]

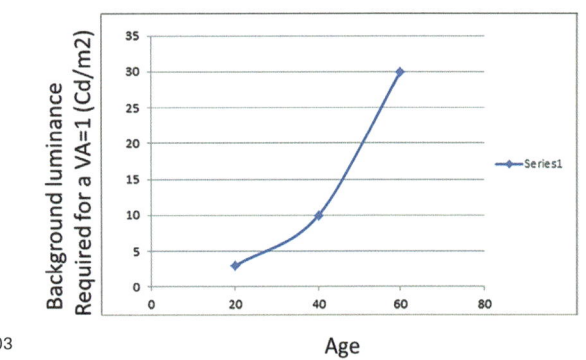

03

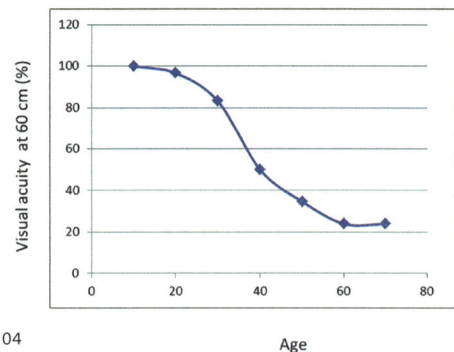

04

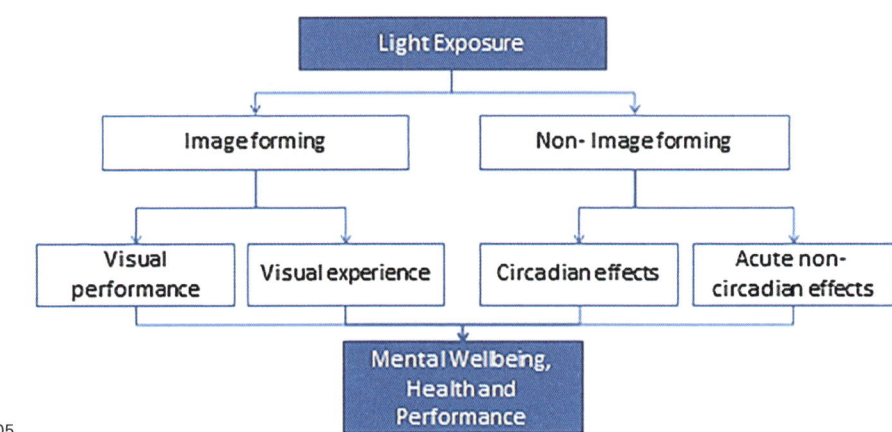

05

LIGHT IS FUNDAMENTALLY IMPORTANT TO OUR LIVES

Besides the image-forming effect of light (vision), the non-image-forming effect of light is fundamental for our health, wellbeing and performance. For instance, bright light during the day enhances concentration and improves mood and sleep, while dimmed light enhances creativity and cooperation. Light is demonstrated to affect the production of melatonin and cortisol, which regulate sleep-wake cycles. When you wake up in the morning, daylight promotes the secretion of cortisol, making you alert. As daylight gets brighter throughout the day, cortisol secretion boosts your performance and creativity. At sunset, when daylight fades, your brain secretes melatonin, a hormone that makes you sleepy. Melatonin levels reach their maximum at midnight, allowing complete rest.

This sleep-wake pattern is extremely important for our bodily mechanisms. As a result, light not only affects sleep and alertness, it also influences our growth, metabolism and immune system. People

03 The need for more light to see the same detail when growing old.
04 The drop in the ability to see small details at a distance of 60 cm when growing older.
05 Schematic representation of the pathways through which lighting influences our mental wellbeing, health and performance.

spend more than 90% of their time indoors, and first indications on the effect of better lighting in these environments is revealed in field studies in hospitals, nursing homes, or education facilities showing improvements in sleep quality and less anxiety,[11] improved sleep-wake cycle, less aggression and more daily activities as well as an improved learning environment, respectively.[12, 13]

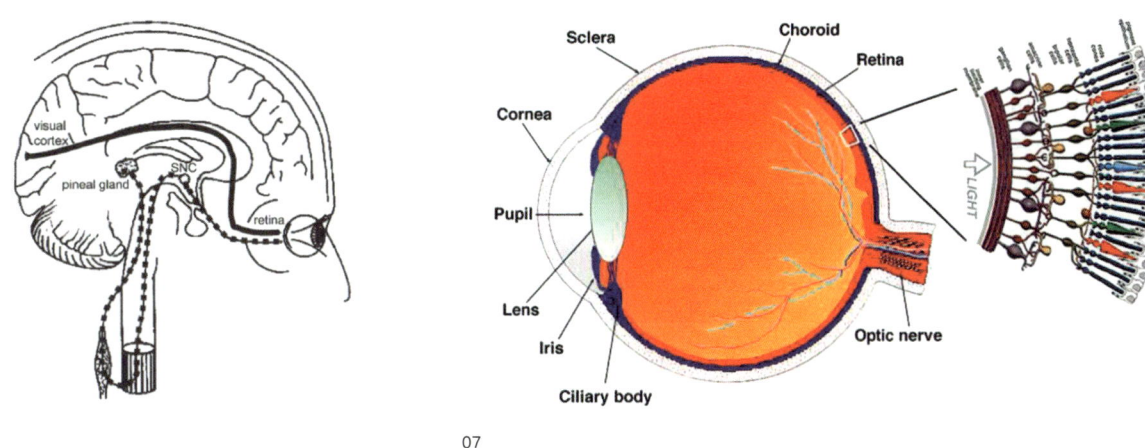

06

07

The science behind the non-imaging impact of light more of less started in 2002 when David Berson et al[14] identified a new photoreceptor residing within a cell type in the retina of the eye. It is called melanopsin and regulates the biological effects of light. When ocular light (light perceived by the eyes) reaches these cells, a complex chemical reaction occurs, producing electrical impulses that are sent via separate nerve pathways to our biological clock, the suprachiasmatic nuclei (SCN). The SCN in turn regulates the circadian (daily) and circannual (seasonal) rhythms of a large variety of bodily processes, such as sleep, and some important hormones, such as melatonin and cortisol, essential for a healthy rest-activity pattern. One speaks of the circadian system, generating the circadian rhythmicity of biological processes. The photoreceptor is most sensitive to blue light (between 446 and 483 nm, with a peak sensitivity at 480 nm).[15, 16]

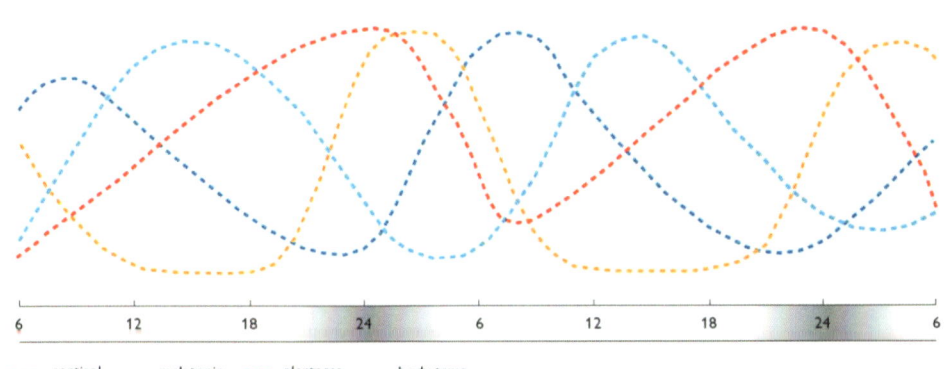

- - - cortisol   - - - melatonin   - - - alertness   - - - body temp.

08

06   Visual (image-forming) and biological (non-image-forming) pathways in the brain: nerve connections between the retina of the eye, with its cones and rods, and the visual cortex (continuous line) and between the retina, with the novel photoreceptor cell, and the SCN and the pineal gland (broken line).
07   Position of the receptors in the retina.
08   2 × 24-hour graph of selected circadian rhythms. This diagram illustrates some typical rhythms in human beings. It shows only a few examples: body temperature, alertness, and the hormones cortisol and melatonin.

THE BENEFITS OF PERSONAL LIGHTING CONTROL IN OFFICE ENVIRONMENTS

These results suggest that working environments would benefit from advanced lighting solutions that accommodate both task needs and different age groups. In 2013, Philips conducted a survey amongst people who tested a desk lamp that allowed them to adjust its light intensity and colour temperature according to personal preference. Although people made very diverse choices in the lamp's settings, 90% or more reported sharper vision, optimum eye comfort, the ability to see smaller details and improved contrast. These results not only illustrate the strong benefits of personal lighting control, it could also be concluded that each person's eyes are unique and that perception of light is very personal.

Light intensity increases visual acuity[17] and reading speed,[18] improves concentration and alertness,[19] and may enhance collaboration.[20] Colour temperature reduces eye fatigue,[21] sleepiness and increases self-reported performance.[22]

The ability to adjust individual workplace lighting conditions according to personal preferences and requirements has been associated with better mood,[23] improved lighting quality ratings, environmental satisfaction,[24] and even higher job satisfaction and higher perceived productivity.[25]

Philips' connected office lighting, based on Power over Ethernet technology, provides personalized lighting for workers. Office employees can adjust the lighting directly above them using an app on their smartphones. This connected lighting system allows office workers to change their light settings depending on their preferences, even in open plan offices. Employees can choose a high light level to boost their energy, or a lower light level to promote creative work. Furthermore, the system can remember their personal preferences.

The LED lights use the same Ethernet cabling as computers, which lowers installation costs. The interaction between ceiling lighting and smartphones is made possible via Visible Light Communication (VLC), a technology developed and patented by Philips that uses light frequency modulation to transmit data. The communication is invisible to the human eye but can be detected by a smartphone camera. Each light point has its own IP address and becomes a device on the network, enabling office workers to control and set their personal lighting preferences in their personal space using their smartphone.

Another important aspect is light distribution, which increases visual comfort and improves the positive biological effects of light, as well as enhancing a room's appearance. And finally, an appealing lighting design also contributes to creating an aesthetically pleasing work environment while providing a high level of psychological and visual comfort for all employees. Research has shown that a positive appraisal of workplace lighting also has an impact on job satisfaction,[26] organizational commitment and employee engagement.[27]

Light is more than the opposite of dark. Light remains a fundamental part of life, whether we spend our time in offices, hospitals, nursing homes, residential areas or education facilities, it will continue to have a significant impact on our health, wellbeing and performance.

As the population continues to age and the way we work keeps changing, the organizations that succeed are likely to be those that seize the opportunities offered by new innovations, including innovations in lighting. By creating lighting that goes beyond just enabling employees to see adequately, organizations will not only see the benefits in people's performance and health, they'll also be likely to see the results in their bottom line.

1 Gensler, 2008
2 Exact percentage may vary across different organizations
3 HermannMiller, 2008
4 Grimley, 2005
5 Sagawa, 2003; Dou, 2011; Adrian 1993; Berman, 1994, 2006; Geerdinck, 2009
6 Steidle, 2011
7 Boyce, 2003
8 Hemphälä, 2013
9 Bitsios et al, 1996
10 Sagawa, 2003

11 Gimenez, 2011
12 Riemersma-van der Lek, 2008
13 Sleegers, 2012; Goven, 2010
14 Berson, 2002
15 Brainard et al, 2001, Thapan et al, 2001.
16 Hankins, 2008
17 Adrian, 1993
18 Mott, 2012; Barkmann, 2010; Fuchs, 2001
19 Steidle, 2010; Hoffmann 2008; Ruger, 2005
20 Galetzka, 2010

21 Dou, 2011
22 Viola, 2008
23 Newsham, 2003, 2009
24 Veitch, 2010
25 Bordass, 1993
26 Charles, 2003
27 Veitch, 2010

Miranda Cheng, mathematician

MATHEMATICAL MOONSHINE

MOONSHINE · INFINITY · MONSTERS · SYMMETRY · NUMBERS · MOONLIGHT

In mathematics the word *moonshine* refers to an unexpected relation between two a priori unrelated mathematical structures: that of very special functions and that of finite groups. The best-known example so far of such a moonshine relation is called *monstrous moonshine*. Concretely, it is expressed in infinitely many equations. The first three of them read

$$1 = 1$$
$$196884 = 1 + 196883$$
$$2149376 = 1 + 196883 + 21296876.$$

The significance of the numbers on the left-hand side of the equations is their appearance in the following special function. The *J-function* assigns a point on an infinitely extending plane (the *complex plane*) to every point on the upper half of a plane. It can be visualized as in [01], where we see a fractal structure of ever-repeating patterns. This infinite pattern, or *symmetry*, can be succinctly expressed in the following equation:

$$J(\tau) = J(\tau + 1) = J(-1/\tau)$$

where τ labels a point on the upper-half plane. While the first part of the equation captures the horizontal repetition in the picture, the second part of the equation explains the presence of the arcs along which the pattern gets mirror-reflected. Moreover, after stipulating the way the function behaves when one goes farther and farther up on the half-plane, these symmetry properties uniquely determine the *J*-function.

Special functions displaying analogous infinite symmetry properties, called *modular forms*, are important elements in number theory. For instance, modular forms played a crucial role in the proof of the celebrated last theorem of Fermat, stating that there is no non-trivial (ie, abc ≠ 0) integer solution to the equation $a^n + b^n = c^n$ when $n>2$, as opposed to the case with $n = 1,2$, where we have solutions such as $3^1 + 4^1 = 7^1$ and $3^2 + 4^2 = 5^2$.

The symmetry property of the *J*-function also guarantees that it can be expressed as an *infinite series*:

$$J(\tau) = 1 \times e^{-2\pi i \tau} + 196884 \times e^{2\pi i \tau} + 21493760 \times e^{4\pi i \tau} + \ldots$$

Note that the first three numbers multiplying different powers of $e^{2\pi i\tau}$ are precisely the numbers that appeared on the left-hand side of the three equations listed at the beginning of this article. More generally, the infinite sequence of numbers given by the above expansion of the *J*-function constitutes the left-hand sides of the infinite sequence of equations.

On the other hand, the origin of the numbers appearing on the right-hand side of the same infinite sequence of equations lies within a totally different branch of mathematics, which bares the name *group theory* and aims to provide us with a rigorous framework in which we can analyze the concept of *symmetries*. Humans have an innate understanding of symmetry. For instance, a toddler's toy familiar to many consists of simple blocks of triangular, square and circular shapes, as well as a box with holes in the corresponding shapes. As the toddler plays with the blocks, rotating them physically and mentally, she learns and subsequently utilizes the growing awareness that these shapes possess distinct symmetries. This learning will help her to identify a hole for the matching shape. In the mathematical language, these symmetries are classified into *groups*: a circle has a continuous symmetry group because it can be rotated at any angle and still remain the same, while a square has a finite, discrete symmetry group because it can only be rotated at an angle that is a multiple of 90° in order for it to remain invariant. More specifically, we say that the symmetry group of the square has $4 \times 2 = 8$ elements, as the square can be rotated at either 0°, 90°, 180° or 270° and it is possible to combine each rotation with a mirror-reflection, all while keeping the square invariant.

Unlike the toddler, mathematicians do not limit themselves to studying symmetries of simple objects. In fact, the objects do not even have to be realizable in the three-dimensional world. By letting our imagination wander into the abstract mathematical world of

01

arbitrary dimensions, mathematicians have been able to classify all (infinitely many) finite groups. All but a handful of them correspond to symmetries of objects that exist in infinitely many dimensions, and as a result these groups come in infinite families. More precisely, there are just 26 outliers not belonging in any infinite family and they are said to be *sporadic*, as they don't seem to follow any regular pattern and correspond to the symmetries of certain exceptional objects. The largest one of these 26 sporadic groups is called the monster group: thanks to its monstrous size it has

$$808017424794512875886459904961710757$$
$$005754368000000000 \sim 8 \times 10^{53}$$

elements, ginormous compared to the eight elements of the symmetry group of a square and comparable to the total number of atoms in the solar system. The existence of the monster group can be understood through its close relation to the symmetry group of a very exceptional 24-dimensional lattice, which describes the most efficient way to pack identical balls in 24 dimensions (imagine if you were an orange vendor in a 24-dimensional street market).

An object of n dimensions which has symmetries described by the group G constitutes an *n-dimensional representation* of G: each element of G can be represented by an $n \times n$ square matrix capturing the way the group element "acts" on this object. Studying such objects upon which G acts is an important part of the content of group theory. For finite groups such as the monster group, any representation can be thought of as built up from *irreducible representations*, which are the basic building blocks that cannot be broken down further. Each finite group G has a finite number of irreducible representations, and the collection of such irreducible representations encodes key properties of the group G.

Now we are finally ready to explain the significance of the numbers appearing on the right-hand side of the infinite sequence of equations we started out with at the beginning of the article: 1, 196883, 21296876 are precisely the dimensions of the three smallest irreducible representations of the monster group! The pattern then repeats itself infinitely many times: the numbers arising from the expansion of the *J*-function into an infinite sequence are always naturally sums of dimensions of irreducible representations of the monster group.

To summarize, this infinite sequence of equations illustrates a surprising and deep connection between the two completely different mathematical worlds, exemplified by the simplest special function – the modular function *J* – on the one hand, and the largest sporadic finite group – the monster group – on the other hand. This is the connection that mathematicians refer to as the "moonshine phenomenon". The effort to understand this monstrous moonshine has proven to be extremely fruitful and led to important progress in different areas of mathematics, and finally to the award of the prestigious Fields Medal in 1998 to the British mathematician Richard Borcherds. Arguably, the discovery and the subsequent understanding of the monstrous moonshine constitute one of the most unexpected and beautiful pages of mathematics in the last century.

After a decade of relative silence, it came as a big surprise when three Japanese string theorists – Tohru Eguchi, Hirosi Ooguri, and Yuji Tachikawa – hinted in 2010 that a new and different type of moonshine might exist. Two years later, another surprising discovery was made by John Duncan, Jeff Harvey and me.

01   Visualization of the
J-function, a fractal structure of
ever-repeating patterns.

$$X_+^{(1)} \cong 0$$

$8n - r^2$	-1	7	15	23	31	39	47
coeff	-1	45	231	770	2277	5796	13915

$$X_2^{(1)} = (-1 + 45q + 231q^2 + 770q^3 + ...) \cdot \left((y - y^{-1}) - (y^2 - y^{-2})q + (y^3 - y^{-3})q^2 \sim \right)$$

$$X_3^{(1)} - E_4 AC \,]$$

$$r = 1, 2$$

$12n - r^2$	-1	8	11	20	23	32	35	44	47	56	59
coeff	-1	10	16	44	55	110	144	280	330	572	704

$$\left(X_4^{(1)} - 3E_4 AB C + 2 E_6 A C\right)$$

$$r = 1, 2, 3$$

$16n - r^2$	-1	7	12	15	23	28	31	39	44	47	55	60	63
coeff	-1	3	8	7	14	24	21	28	56	43	59	112	94

$$\left(X_5^{(1)} - 3E_4^2 A^3 C + 8 E_6 A^3 BC - 6 E_4 A B^2 C\right)$$

$20n - r^2$	-1	4	11	16	19	24	31	36	39	44	51	56	59	64	71
coeff	-1	1	4	5	4	6	11	15	9	10	24	26	20	25	45

02

Inspired by a page from a notebook [02] of the renowned number theorist Don Zagier that he shared at a conference in Zürich in 2011, we discovered that there are in fact 23 such instances of new moonshine, which we named *umbral moonshine*. Moreover, these 23 cases are controlled by the 23 special lattices in 24 dimensions – the so-called Niemeier lattices – which are again closely related to the optimal way an imaginary orange vendor can pack his/her 24-dimensional oranges!

The new moonshine is reminiscent of but different to monstrous moonshine. This time other sporadic groups, different to the monster group, are involved. On the other hand, the role of the modular *J*-function is replaced by the so-called *mock modular forms* – itself a subject full of mysteries and surprises and whose study was initiated by the legendary mathematical genius Srinivasa Ramanujan in 1920, the last year of his very short life. A special feature of this type of function is that each comes with its own *shadow* function, which led us to adopt the adjective "umbral" when naming the new moonshine. At the time of writing, mathematicians around the world (including myself) are still struggling to gain an understanding of this new moonshine. The puzzle is again: why do the numbers coming from two totally different worlds coincide with each other?

Mathematics has proven to be "unreasonably effective" for describing our world, as the physics Nobel Prize Laureate Eugene Wigner put it.

Why this is the case is a deep question that thinkers of various kinds have been debating passionately since the time of Pythagoras. Is mathematics discovered or invented? Does mathematics exist in some abstract world, with humans merely discovering its truths? Or is mathematics a human invention, specific to the brain of our species and how it perceives and interprets reality? While Wigner eloquently advocated the former view in his 1960 essay "The Unreasonable Effectiveness of Mathematics in the Natural Sciences", the German mathematician Leopold Kronecker (in)famously claimed in 1893, "God made the integers, all else is the work of man" (*Die ganzen Zahlen hat der liebe Gott gemacht, alles andere ist Menschenwerk*).

02   Page from Don Zagier's notebook. Photo taken by Miranda Cheng in 2011. The notebook no longer exists.

This nearly metaphysical debate will undoubtedly continue for a long while still. Here I'd like to share an insignificant personal observation. Seeing pages and pages of numbers churned out from laborious analysis of special functions on the one hand, and then unexpectedly encountering again and again the exact same numbers on a journey through the exotic landscape of a higher-dimension world with its exceptional symmetries elicited a (very human) response: I marvelled at the mathematical truth, so much larger than us, yet is as real as the moon that shines.

Steven Rose, Oliver Pike, Felix Mackenroth, Ed Hill, physicists

MATTER FROM LIGHT

ENERGY MASS ANTIMATTER PHOTONS EXPERIMENT

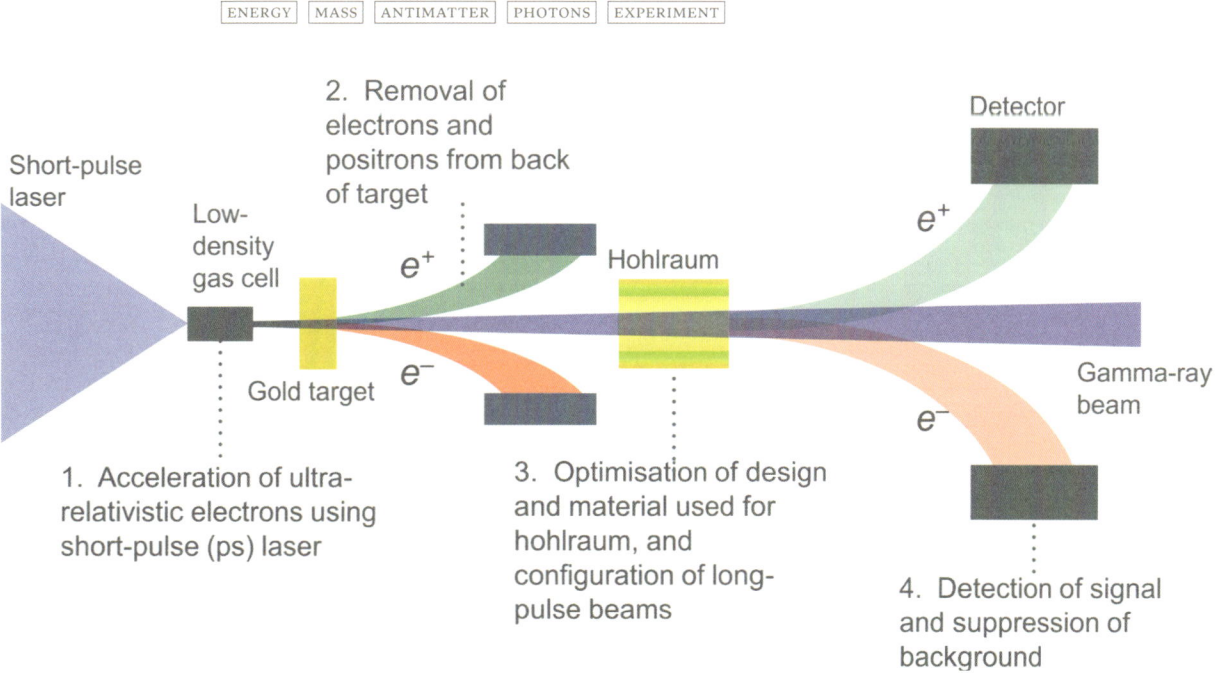

Short-pulse laser

Low-density gas cell

2. Removal of electrons and positrons from back of target

Gold target

e^+

e^-

Hohlraum

Detector

e^+

e^-

Gamma-ray beam

1. Acceleration of ultra-relativistic electrons using short-pulse (ps) laser

3. Optimisation of design and material used for hohlraum, and configuration of long-pulse beams

4. Detection of signal and suppression of background

The aim of this experiment is to create matter directly from light. This will be a pure demonstration of Einstein's famous equation that relates energy and mass: $E = mc^2$. It has never been done before using the simplest possible process where two "particles" of light (photons) come together and create matter (and antimatter) directly.

01   e^- is an electron and e^+ is a positron; they are matter and anti-matter (respectively) – so when we produce both e^- and e^+ we produce matter and anti-matter in equal amounts.

The process was first predicted in 1934 by two famous physicists, Gregory Breit and John Wheeler, using the then new theory of the interaction between light and matter known as quantum electrodynamics (QED). Whilst every other fundamental prediction of QED has been demonstrated experimentally, the "two-photon Breit-Wheeler process" has never been seen. Our work proposes a way to see it in the laboratory using equipment that already exists. That equipment involves high-power lasers. There are only a few laboratories in the world that have this equipment and we are now trying to get the experiment done.

One point worth making is that although we talk about two photons of light coming together to produce matter and antimatter, it is not really light as most people know it. The light is not in the visible region of the spectrum; indeed one of the photons that we will use has about 1000 times and the other has 1000 000 000 times the energy of visible light photons.

Tadao Ando, architect

PERPETUAL QUEST FOR THE LIGHT

INTERPLAY   SHADE   STILLNESS   DARKNESS   SPACE   CONTRAST

I know two different kinds of light. One is the dim light existing in the darkness, as if embraced by the darkness and blending into the darkness. The other is the vibrant light slashing across the space, claiming its own existence in strong contrast to the darkness. Japanese traditional houses first to spring to mind as spaces of the kind of dim light: the latter kind of vibrant light can be found in European masonry cathedrals. In Japanese traditional architecture the light indirectly comes from below: eaves and "shoji" (paper screens) obscure direct sunlight, and the light that is reflected from the garden and "engawa" (wooden veranda) gently enfolds the people in the space. On the other hand, the light in European architecture is more direct and dynamic. The aim of architects to control light as an element of architecture and structure is apparent. In the Pantheon in Rome for instance, the cylinder-shaped light directly falls from the dome zenith to the marble slabs on the floor. This light breathes life into one of the greatest architectural spaces, which contains simplicity and self-consistency. This dynamic between space and light did not originally exist in Japan.

The history of European architecture can be interpreted as the development of techniques for characterizing and bringing light into buildings. Walking across Europe to explore the world of Western architecture, what impressed me most was the light itself and how strongly it characterized different spaces.

However, while I try to make use of light dramatically, like in the Pantheon, I also wish to nurture space with the Japanese dim light, which blurs into the darkness.

01–02 Kochino House, 1981–84.

01–02

03–04

I cannot describe in words the spatial awareness created by shadows, and the experience of the essence of space *as* the darkness, but it is deeply ingrained in my body. Because of the experience accumulated over time and through daily life, that particular sense of the darkness penetrated more deeply and strongly than I ever would have thought.

Through the experience of both the Japanese cultural space of the light characterized by shadows, and the Western rational and three-dimensional space of light, I have lived as an architect seeking my original space for the light and the darkness. It is not simply an either-or choice, but the process of a never-ending conflict concerning those two contradictory perspectives and realizing them both in my own architecture.

"Row House in Sumiyoshi" (1976) opens its central part, one third of the whole concrete-structure box, to the sky as a courtyard. "Koshino House" (1981, 1984) was designed to express the sequence of the space as a drama of light. If my work has any recurrent themes, one is definitely the pursuit of light. The reason why I insist on exposed concrete is because it purifies the architectural space to the light and to the flow of air. In other words, it is the most suitable way to realize my idea of the purest space.

"Church of the Light" (1989) is typical of my work in that it is genuinely concentrated on the expression of light. A ray of light coming through the cross-shaped slit on the front façade symbolically illuminates the rough-boarded floor in the darkness of the small concrete box, which excludes any additional elements. The cross-shaped light is the essence of the church. Light lets us perceive textures of materials that cover the space, and simultaneously recognizes the existence of the space itself.

I want the interplay between the space and the light to speak to the hearts of people through the five senses. I keep fighting with the light in architecture, and and continue on my quest for my ideal space of light.

03–04 Church of the light, 1989.

Jan and Tim Edler (realities:united),
artists, architects

LIGHT ABUSE

COUNTERINTUITIVE MESSAGE MODIFICATION
INTERACTION SPACE

01–02

Light, artificial and natural, has always played an important role in our installations. Not because we want to prioritize light over other materials, or because we are passionate about designs involving light and shade, but rather for pragmatic reasons. Because light, or more accurately, the technologies that are currently available for the creation of light, seem to speak to the specific ideas embodied in our works. To put it another way: if another medium had been available that had better suited the realization of our goals, we might well have used it. This design opportunism seems in retrospect to have led to our most surprising applications and discoveries.

One thing we are known for is how we incorporate light to a building's façade to share information. The idea of the media façade is by no means a new one. Artists and architects such as László Moholy-Nagy and Oscar Nitzschke experimented with the architectural integration of moving images as early as the 1920s. Nitzschke's 1926 "Maison de la Publicité" is one of the best-known examples. And the idea of façade-sized urban screens has been present in popular culture at least since Ridley Scott's dystopian science fiction film *Blade Runner* (1982).

We never intended to convert building façades into media displays as such, or to apply the established technical formats propagated by the display industry onto architecture. All the projects we developed in that field are closely linked to a specific context and architecture and follow our greater interest in and research on the creation of dynamic, ie, changeable architectures.[1] The latest culmination of this research is our communicative façade design for the Contemporary Art Centre in Córdoba [01–03], which we developed in collaboration with Nieto Sobejano architects. The façade consists of elements that function as individual pixels designed to reflect the building's complex inner geometric structure, and operates as a display system with an extremely low resolution arranged in different grids of varying density, size and shape.

In parallel to the projects for changeable building shells, we started to think about the artistic and communicative potential of the

03

01–03 C4, communicative lighting facade, Espacio de Creación Artística Contemporánea, Córdoba, Spain, 2013.

04

05–06

technical infrastructure already present in buildings. The research project "NIX"[04–06], which has been running since 2003, takes this basic approach to the artistic use of available lighting and applies it on the scale of fully glazed, high-rise office buildings and even groups of buildings. After office hours the illuminated ceilings would be centrally orchestrated and transform the entire urban skyline into an abstract work of art. The architectural critics Ilka and Andreas Ruby have referred to this intervention as a "deep-tissue massage of architecture".[2] The project title "NIX", a German slang word meaning "nothing", plays on the notion that we don't add extra lights, form or material. All we are doing is taking over the orchestration of the lighting system already in place. This vision is still a work in progress.

This "anyhow technology" (using what is already there) is also at the heart of the light-installation for the "Toni Areal" [07–08] in Zurich (2014), which challenges perceptions of space. Through the spatial and organizational fusion of Zurich University of Applied Sciences and Zurich University of the Arts, a new super-institution was created in the former Toni dairy, remodelled and extended by the Swiss architectural design firm EM2N.

The internal access area of more than 6000 m² doubles up as a student exhibition space that extends throughout the building. This semi-public space is a place for discussion and debate based on diversity and competing interests.

Our intervention was achieved without introducing additional technological components or design elements. We moved and reconfigured the already existing lighting infrastructure. The off-the-shelf fluorescent tubes are concentrated asymmetrically in massive clusters in the different sections of the space. The arrangement of the lighting produces an uneven brightness, creating a dynamic field of tension in how the space is perceived and experienced. This energetic division of the space is confusingly indifferent to the architecture. Our erratic, "wild" lighting system does not obey the usual strategies and rules of lighting design. It doesn't follow any technical grid arrangement and has no relation to other characteristics usually associated and expected from lighting designs. Nor does the conglomeration of lamps result in an unambiguous light sculpture or serve as a dramaturgical accentuation of the space or the architecture.

04–06 NIX, research on the artistic potential of synchronized lighting systems in high-rise buildings, since 2003.
07–08 Toni Areal, artistic light installation for the ZHAW and ZHDK in Zurich, Switzerland, 2014.

Instead, this "bad lighting" installation seeks to serve as a catalyst that fosters and demands the users' appropriation of these spaces. Students who exhibit their work will not find surfaces optimized for exhibition operations, but rather a place that, in the process of appropriation, provokes interaction with and possibly changes the given conditions. All modifications and misapplications are very welcome!

1   Eg, BIX, communicative display skin for the Kunsthaus, Graz, Austria, 2003; SPOTS, temporary light and media installation, Potsdamer Platz, Berlin, 2005; CRYSTAL MESH, ornamental and granulated light and media facade, Iluma, Singapore.
2   Ilka Ruby and Andreas Ruby, "Spatial Communication, A New Quality of Media Architecture throughout the Work of realities:united" in *realities:united featuring*, edited by Florian Heilmeyer, Ruby Press (2010)

07

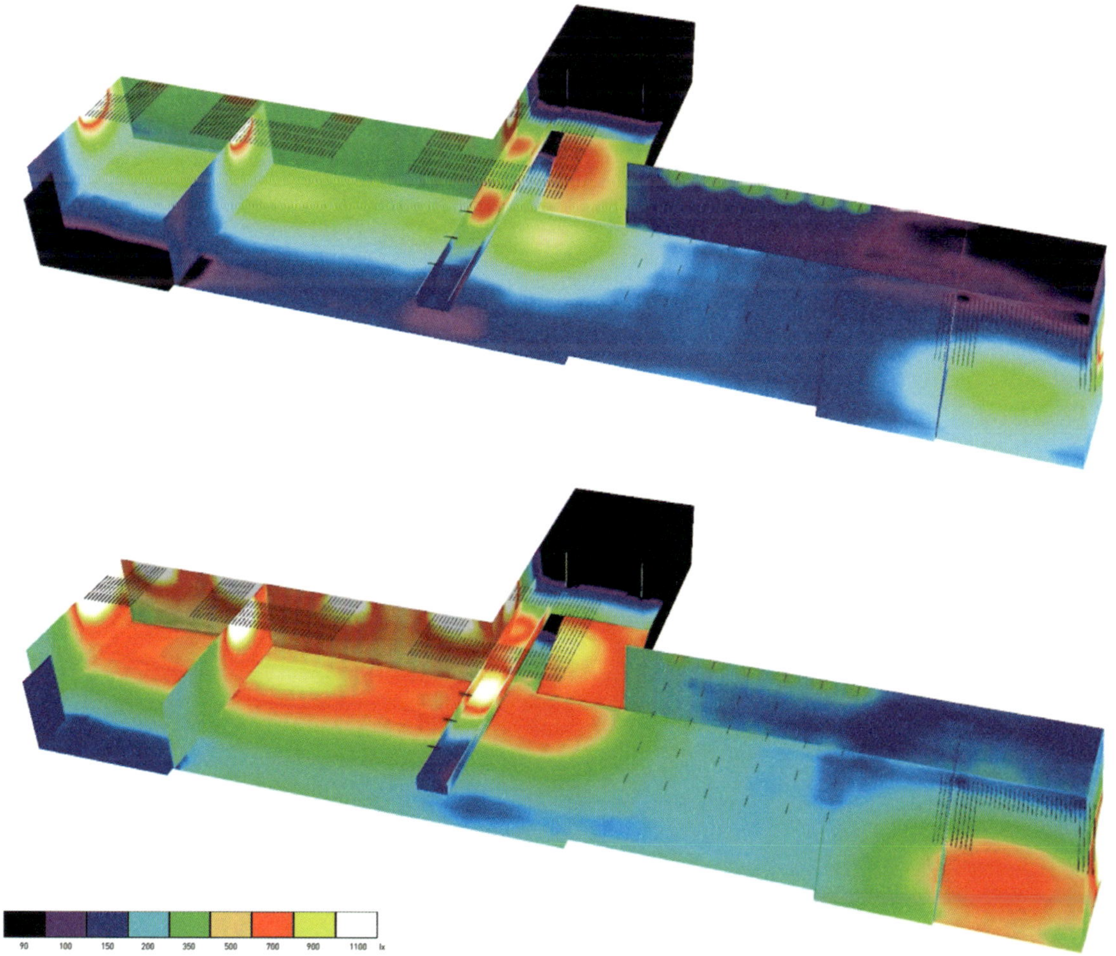

08

Katy Evans-Bush, poet

POEM IN WHICH FASHION IS A VACUUM

VAVAVOOM   SEDUCTION   VOLUPTUOUS   INTERFERENCE

Vanta Black, fifties chanteuse
was a killer in her little black dress.
Most girls wore feathers, or silk of chartreuse,
but she simply sucked out the light to impress.

Iván Navarro, artist

TRAFFIC

UNDERCURRENT   TORTURE   POWER

Traffic lights, metal tubes and electric energy, 2014

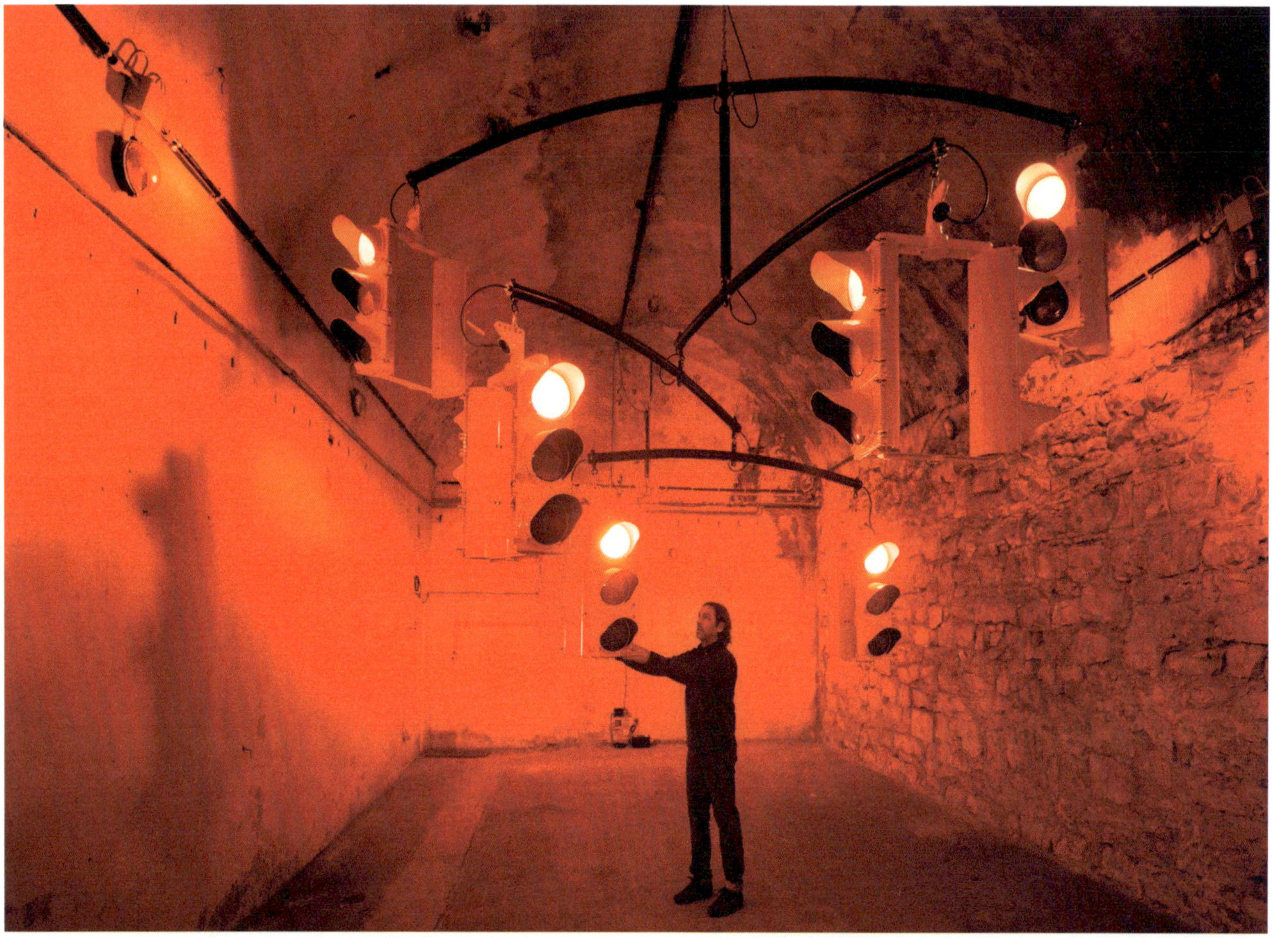

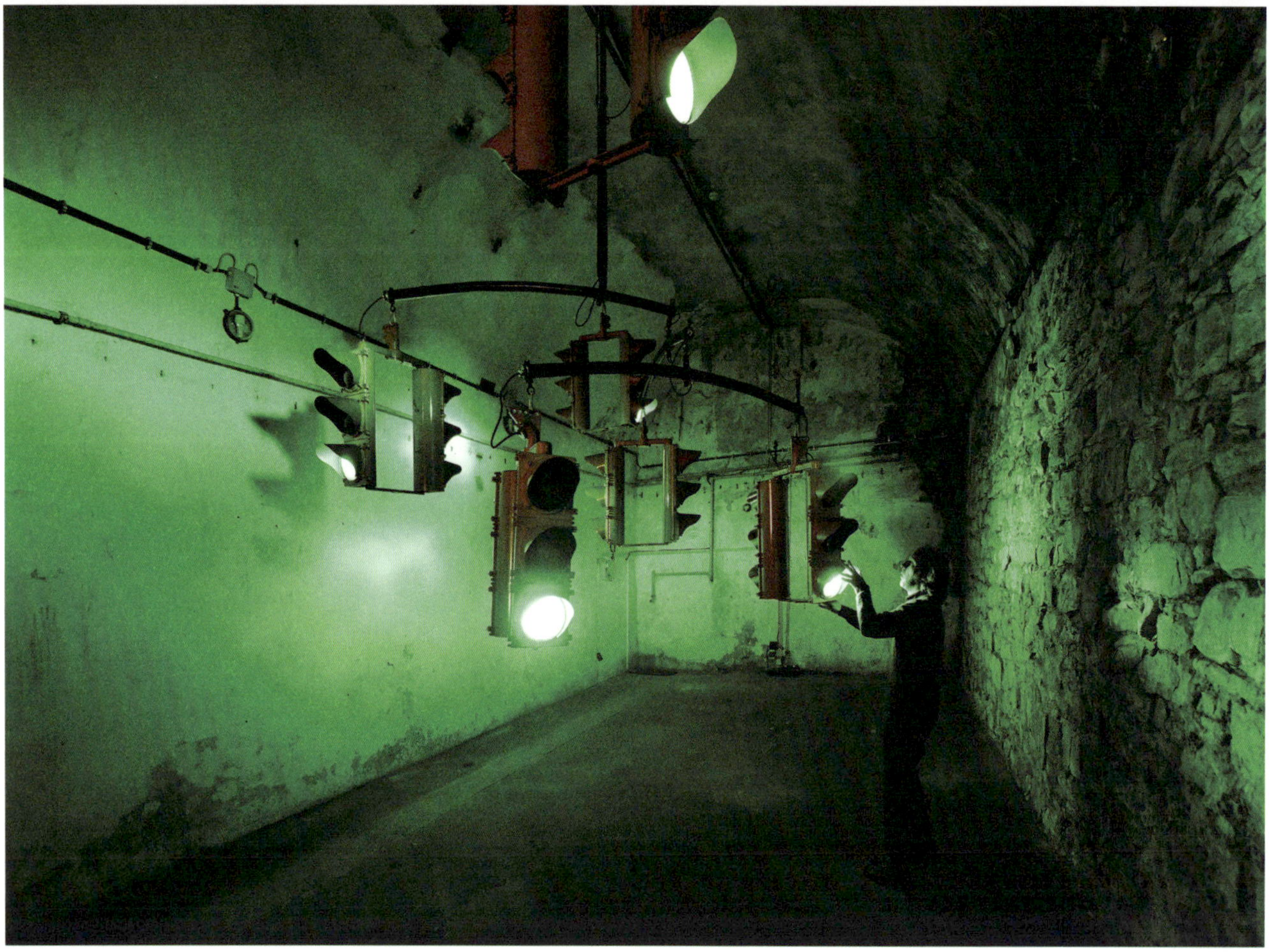

Marta Brusasca, physicist

HYPERSPECTRAL IMAGING

EYE-LIGHT CAMERA INVISIBILITY TOMATO FORENSIC

Our body is a multi-sensor platform. We experience the world by seeing, tasting, hearing, touching and smelling the environment around us, acts which are a mixture of physical and chemical analysis. What we feel with one sense is usually cross-checked by the others: the eyes see a rotten peach, the nose smells the decomposition process caused by bacteria, and the tongue tastes the fermentation of the sugars as the peach putrefies.

But there are limits to our senses, especially to our perception of light. The human eye can see only in the visible range 400 nm to 700 nm. Technology enhances our limited senses with hyperspectral imaging, assisting our way of seeing with a camera system that can perceive spatial coordinates and many more spectral features at once.

Hyperspectral sensors look at objects using a vast portion of the electromagnetic spectrum, decomposing it into many more wavelengths than the red, blue and green that our eyes can see. Information is collected as a set of images at different wavelengths, usually in the ultraviolet, visible and infrared spectral bands. These images are then combined and form a three-dimensional hyperspectral data cube (x,y,λ) for processing and analysis.

Standard multispectral image classification techniques were generally developed to classify samples into broad categories with the purpose of finding objects, identifying materials or detecting differences. Hyperspectral imaging provides an opportunity for more detailed image analysis and is used in different fields as an integrated technique to fuse physical and chemical data.

One example is the forensic hyperspectral camera that detects blood at a crime scene: bloodstains are recognized and dated without any chemical analysis in real time.

Food inspection is another interesting application of this technology. Sampling and destructive analyzes are expensive and time consuming, while light can give

01

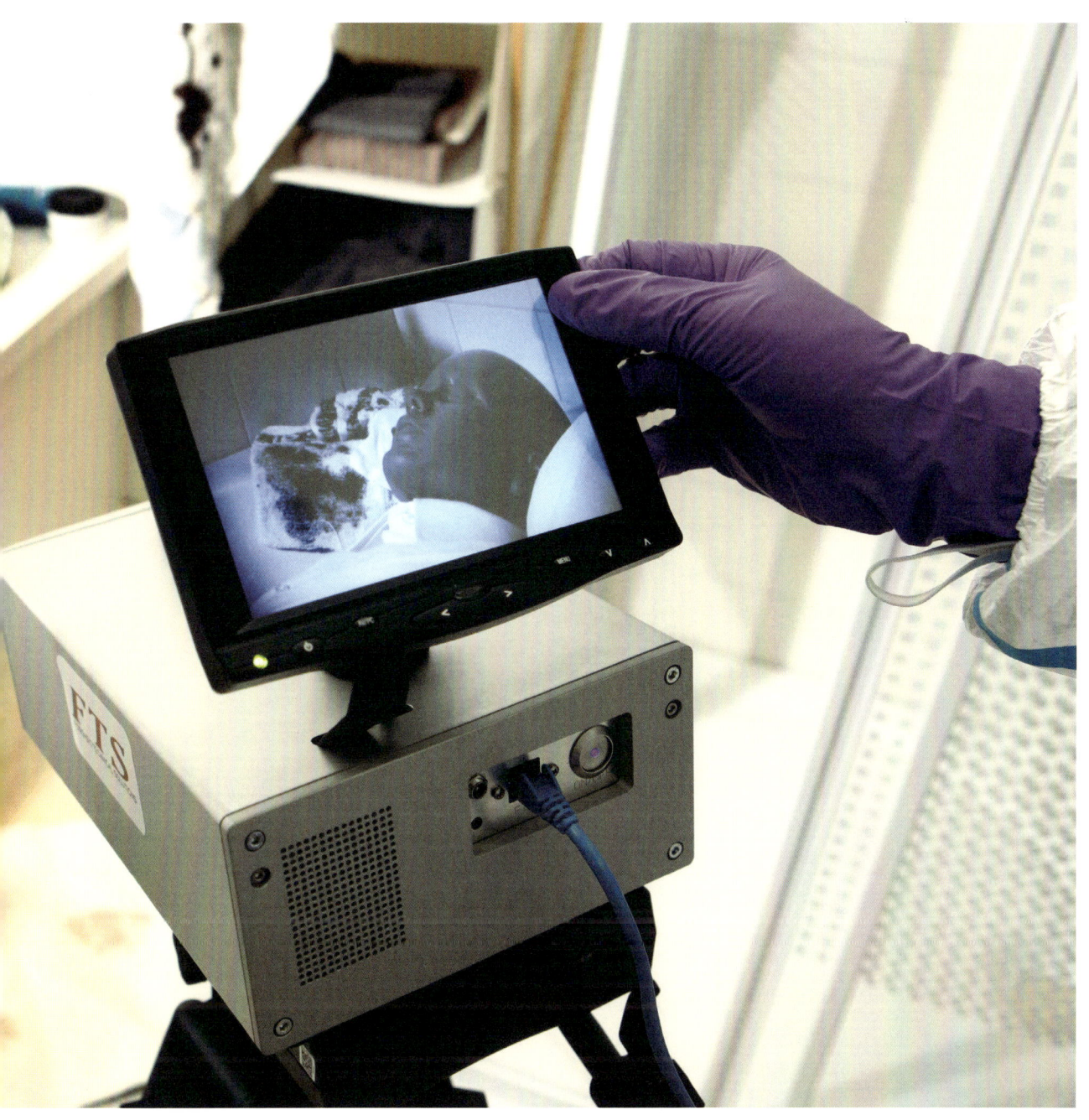

us a deep insight into the food status in a fast and safe way. Adding intelligence to the normal visual inspection is possible with hyperspectral cameras that merge the optical investigation with the wet chemical laboratory analysis. The next generation of advanced inspection systems offers non-contact, non-destructive and (even more attractive) real-time methods to safely check food quality. Contaminants, physical and/or chemical properties, like moisture content, stage of ripeness or freshness status can be measured. The fingerprint of a particular substance can be detected in the recorded spectra because emission or absorption peaks indicate the presence of chemical components. These substances

01 The hyperspectral imaging camera that cosine developed for forensic applications.

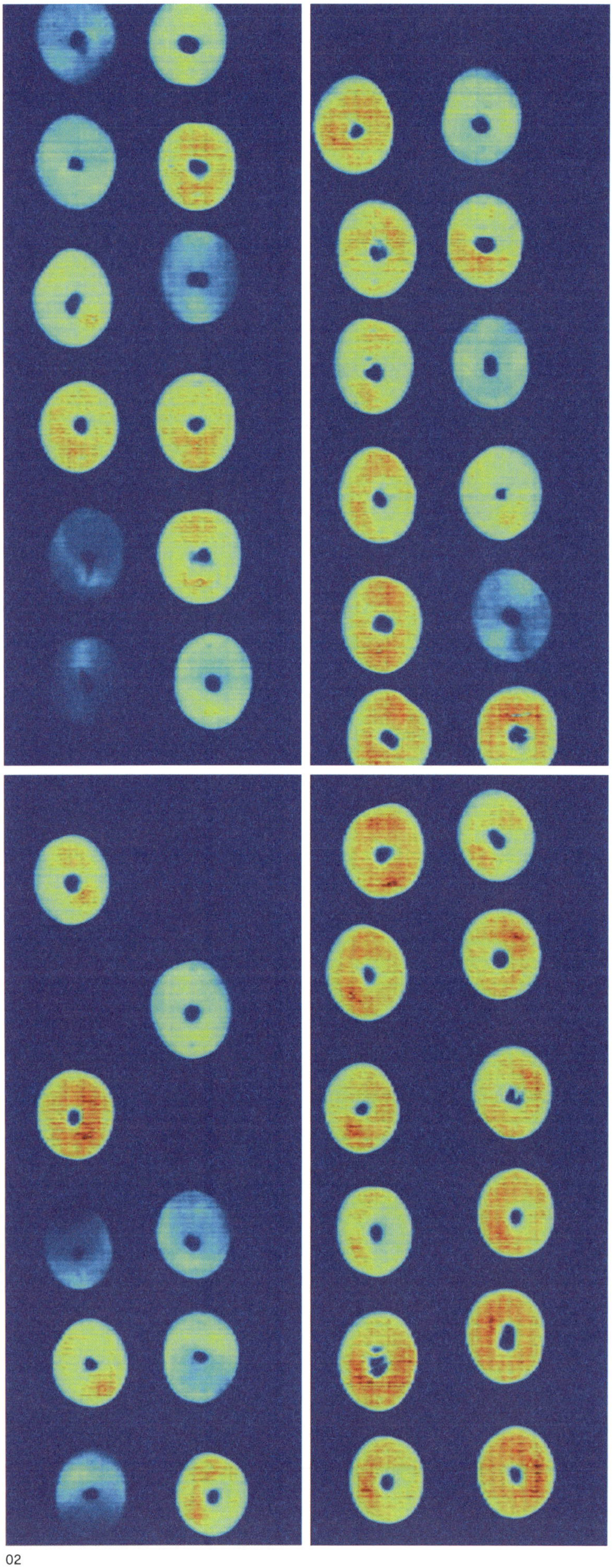

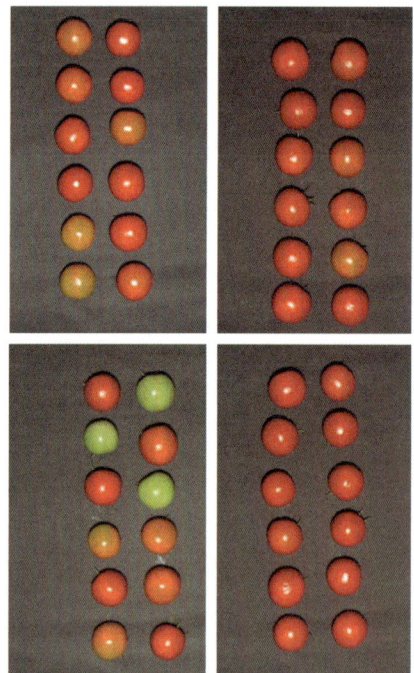

02

03

02–03 Hyperspectral images
of tomatoes and RGB: stage of
ripeness can be measured.
04   CONDI food hyper-
spectral camera.

can be highlighted directly on the hyperspectral picture representing the sample. This provides, not only a point-like measurement, but also a real-time image of an object, showing rotten parts of an apple or chemicals spread on the skin of a mango, for example. The sample can be illuminated with different light sources: UV light detects fluorescence signals, while halogen lights detect colour or thermal emission in the visible and near-infrared bands. This enhanced type of observation enables the use of monitoring techniques to identify and predict safety issues.

Objective evaluations can be used to grade the products. What an expert in the field considered a subjective decision can now be evaluated through optical measurement systems. Training data sets are used to create databases of information correlated to the product. The red colour detected by the hyperspectral camera indicating the ripened tomato can now be distinguished from all the other reds that do not indicate ripeness.

05

Another important development is the application of the hyperspectral camera to fish freshness. The expert's freshness evaluation can now be verified using data taken from the blood in the gills.

Many parameters can be measured with hyperspectral cameras: ripeness or bruises on fruits and vegetables, authenticity of formula milk, presence of contaminants or undesired contents, and, for example, the origin of olives (traceability), fish freshness or the varying intensity of roasted coffee beans.

Much in the same way that the authenticity and provenance of an artwork can be determined (hyperspectral imaging is also used for art fraud), the fingerprint of a product can be identified using hyperspectral imagers. These fingerprints correlate to chemical elements present in a sample and are a measure of authenticity and traceability.

Hyperspectral imaging goes beyond our senses. The technology allows us to see the invisible, which often hides a wealth of information. From the origin of the universe to the freshness of a peach, light is a probe with which to investigate our world.

05   IR picture of roasted coffee beans: the burnt one can be easily detected.

Paul Struik, crop physiologist

HARVESTING THE SUN

PHOTOSYNTHESIS   PHOTORECEPTORS   PLANT GROWTH   ENERGY   AGRICULTURE

Light is vital for life on earth. Light is an almost infinite source of physical energy in the form of photons, discrete bundles of energy or light quanta. However, only certain life forms (plants, algae, cyanobacteria, etc) can actually capture this form of energy through special light-harvesting complexes and transfer that energy to special reaction centres. These organisms convert this energy into chemical energy, which can then be used for biochemical processes. Through the very complex metabolism of these organisms they are able to use this energy to synthesize compounds they need to function. The raw materials for these synthesis processes consist of carbon dioxide from the air, water and nitrogen, and other nutrients from the soil. Plants can produce hundreds of thousands of different compounds. The most important, primary process in biochemistry is the fixation of carbon catalyzed by the enzyme rubisco. This is the most abundant enzyme in the world, but not very well designed as it also catalyzes the reverse process, resulting in a great loss of efficiency. The process of capturing energy from photons of light, charging the bio-battery in reaction centres, the creation of chemical energy, and the fixation of carbon to produce assimilates using that energy is called photosynthesis. In plants, it takes place in special, small organelles (cell factories) in the leaves, called chloroplasts. As carbon dioxide is sourced from the ambient air, leaves require an efficient gas exchange and transport system. An important waste product of the process is oxygen, which then becomes available to respire all that fixed carbon again at a later stage, for example, when a plant has died. Gas exchange is accompanied with transpiration. Therefore oxygen production, carbon fixation and water loss are inseparable.

Because of this ability to photosynthesize, plants are at the beginning of the food chain. The organic world, mainly consisting of carbon, oxygen, nitrogen and hydrogen, is based on this phenomenon. Think about that when the sun rises in the morning! Another avalanche of photons; if only we could make better use of them!

Mankind has created a complex system of agriculture to harvest the sun by domesticating plants and animals that are efficient in producing the products that people require. Domesticated plants and animals also profited: these species have been made into very successful colonizers of planet Earth and are currently dominant in areas far beyond their centres of origin and in habitats in which they would not survive without the help of humans.

Because of the nature of their light-harvesting systems, plants can only use a certain fraction of the light spectrum for photosynthesis. A lot of light is not photosynthetically active. For example, the green part of the light spectrum cannot be absorbed and is therefore reflected. That is why plants are green. Photosynthesis can be limited when there is not enough light to run the energy production at a high rate. It can also be limited because the activity of the enzyme is not high enough, for example because there is not enough carbon dioxide or (when there is plenty of light and carbon to fix) because the rate at which the photosynthates are used is limited (for example by a limitation of triose phosphate utilization). Moreover, optimal light conditions and the presence of healthy, young leaves do not always coincide, which is another major reason for inefficiency. In general, land plants are not particularly efficient in converting solar energy into biomass. Typical solar energy conversion efficiency by annual crops is only 1–4% on the basis of incoming global solar radiation across a full growing season.

There are many reasons why solar energy is lost. But plants usually could not care less. Plants have not evolved to maximize energy conversion efficiency; they are there to produce a next generation, thus passing on their genes to offspring. They only need to produce enough and in a timely manner to make that reproduction happen, in competition with neighbours of the same or other species, and before being eaten by herbivores, burnt by fire or killed by frost, heat, drought or other abiotic and biotic stresses. Some desert plants are known to have a very high rate of photosynthesis, much higher than our most efficient crop plants. These desert plants need to complete their life cycle very fast on the very little rain that is available during a short period of time.

For plants, light can also be a nuisance or even a danger. Plants cannot hide from light as they are sessile and can become overheated or overexcited when the intensity of photons becomes too high or when there is not enough water to cool them through transpiration or when temperatures are too low to let the photochemical and biochemical machinery run properly. Fortunately, plants have many protective mechanisms to safeguard their light-harvesting systems from overexposure, for example heat dissipation and different forms of nonphotochemical quenching. But letting the whole machinery of photosynthesis run efficiently is a balancing act for a plant, certainly when considering the dynamic nature of the availability of light in canopies. Just think about sun flecks, creating continuously changing conditions of high light intensity alternating with shade in a forest, a crop or any vegetation for that matter.

Photosynthesis is a very complex but also very informative process. Measuring in great detail how a plant responds to light can tell us a lot about the plant. Not only about how much biomass the plant produces, but also about the plant's wellbeing: in this respect photosynthesis can serve as a thermometer of the plant. If only we could redesign and engineer the process in such a way that its efficiency is doubled, especially now that we have higher ambient carbon dioxide concentrations. It would enable us to enhance agricultural productivity enormously, provided of course we have enough water to keep the crops functional.

Light is more than an energy source. The sessile nature of plants makes it necessary for them to receive signals from the environment, to interpret these signals wisely, to communicate that information to other plant parts not exposed to the signal, to make predictions about future opportunities of growth and development on the basis of those signals, and to make decisions based on cost-benefit analyses about how the rest of the life cycle should be organized in order to enhance fecundity. Signal perception and interpretation, communication and organization are essential for a plant to survive. We are only at the very beginning of understanding how plants do that.

In addition to using light as an energy source, a plant uses light to "read" its environment. This is what makes the role of light in the life of plants so exceptional, fascinating and exciting.

What does light do to a plant in addition to providing energy? First of all there is proof that the rate of photosynthesis has a direct impact on gene expression in plants. Light thus directly determines which part of its genetic potential is activated.

MOREOVER:

— Light can work as an alarm clock. A single flash of light can act as a wake-up call to rouse a seed from its darkness-induced dormancy and to start germinating.

— Light (especially certain wavelengths, for which plants have special receptors) can work as a timer. The plant assesses what time of day it is, based on the sun's altitude and azimuth, and behaves accordingly: sunflower heads moving with the sun, for instance. And light helps plants set their circadian rhythms. Based on day length in combination with cold or heat periods, short-day plants and long-day plants can also adjust their phenology to the environmental conditions of the seasons.

— Light tells the plant in what direction to move (phototaxis) or grow (phototropism). Plants can determine the location of the light source, what is up and what is down in response to the directional component of light perceived by a blue-light receptor. This also helps the plant to remain properly anchored and to position certain organs to optimize exposure to light (eg, through torsion in certain plants) or seeking shelter from too much light. A special case of the latter are compass plants, which orient their leaves towards specific compass points in order to capture the sun during certain parts of the day while avoiding direct exposure when the sunlight is too intense.

— Light determines the shape of the plant, such as leaf size and thickness, and stem length, through photomorphogenesis, based on the activity of different types of photoreceptors.

— Ratios of certain wavelengths (red: far-red ratios) inform the seed or plant whether it has neighbours potentially competing for light or other resources. This observation helps a seed to "decide" whether germination is timely or will trigger plants into developing shade-avoidance strategies. If there is plenty of space seedlings or plants might start to tiller or to branch. If it really is getting too crowded self-thinning might occur,

The processes described above are only possible because plants have very diverse photoreceptors, including chlorophyll, the blue-light receptors phototropin and cryptochrome, the red-and far-red-absorbing photoreceptor phytochrome and UVR8 for UV-B light reception. Given the fact that certain wavelengths produce the best energy capture, that other wavelengths have an impact on photomorphogenesis, and that photoperiod affects phenology, one can influence the growth and development of plants under semi-controlled conditions in a very energy-efficient way. In glasshouse horticulture or in completely controlled production systems without natural light we can shape the plant, make it grow efficiently and make it as nutritious and healthy as possible. Once we fully understand the impact of light and can produce our own efficient light source that triggers exactly those processes, we can then create our ideal plant under completely artificial but well-controlled conditions. In the production of vegetables this is already feasible and ongoing in technologically advanced production systems. It has already been done in space stations and will be possible in other extraterrestrial biospheres in the future. For my research group the biggest challenge is to create a thorough understanding of energy capture through photosynthesis at the subcellular level. If we can scale up this research into an overarching understanding of all the effects light has on plant and crop systems we could use this knowledge to increase light use efficiency and genetically improve plant and crop photosynthesis. The plants we will produce in the future might look odd but they will be efficient in the use of resources. Our lettuce may never look the same.

meaning that some individuals will sacrifice themselves to allow others more space and more resources.

— Light informs the seed, bulb or tuber how deep it is buried, so that it avoids suicidal germination or fatal sprouting.
— Light and light quality tell the perennial plant to become dormant when periods of harsh growing conditions are nigh.
— Light helps the plant to allocate its resources in such a way that they are used efficiently. For example, often the distribution of nitrogen, necessary for light interception and photosynthesis, among other processes, matches the distribution of light in a canopy.

Hamid Ismailov, writer

A CORPSE

ETERNAL   MIRAGE   SHADOW

Let him who gives me a shadow not hold me.
You know the breadth of a star
is not equal to the embrace of the ray.

Let me go, blue holy light,
my shadow is in torment on the black earth.
Am I drunk, or is my road drunk?

The snow flows, the earth is white and black.
The word 'I' is a wanderer like I,
you are eternal as an icy, cracked puddle.

Did we trip over our shadow
or did the mirage melt in the icy pupil—
a roof, holding up a lamp,
    when the house moved.

Cuppetelli • Mendoza, artists

VARIOUS INTERFERENCES

WAVE INTERFERENCE

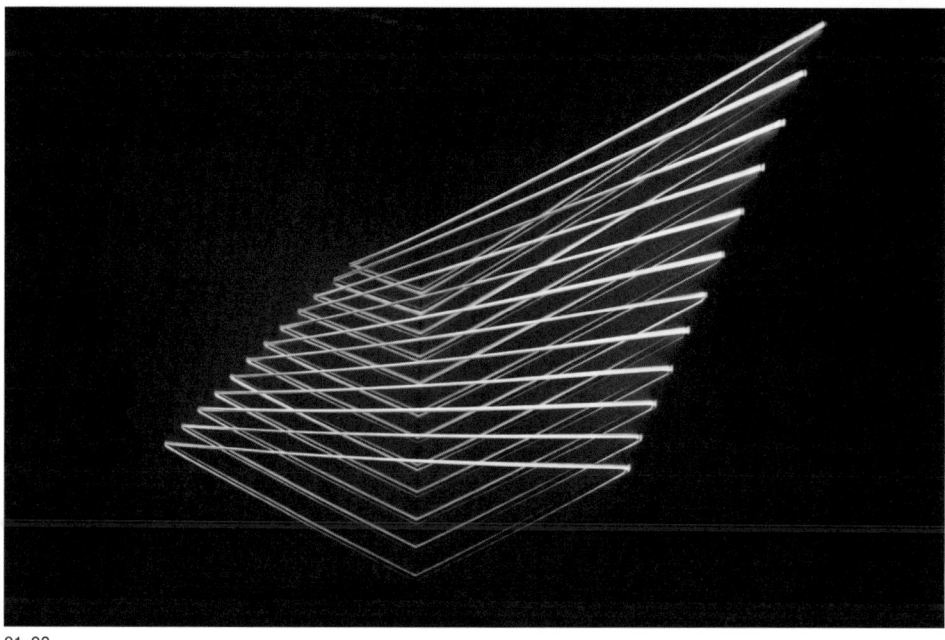

01   *Threaded Interface*. MDF,
elastic cord, custom software,
computer, video projectors, video
cameras, 2012.
02   *Nervous Structure 5*.
Site-specific interactive installation,
spandex, video projector, video camera,
computer, custom software, 2011.

03; 05 *Notional Field*. Site-specific interactive installation, MDF, elastic cord, video projectors, video cameras, computer, custom software, 2012–2013.
04; 06–07 *Transposition*. Site-specific interactive installation, MDF, elastic cord, video projectors, video camera, speakers, subwoofer, computers, custom software. Sound composition and programming by Peter Segerstrom. Denver Art Museum, 2013.

03–04

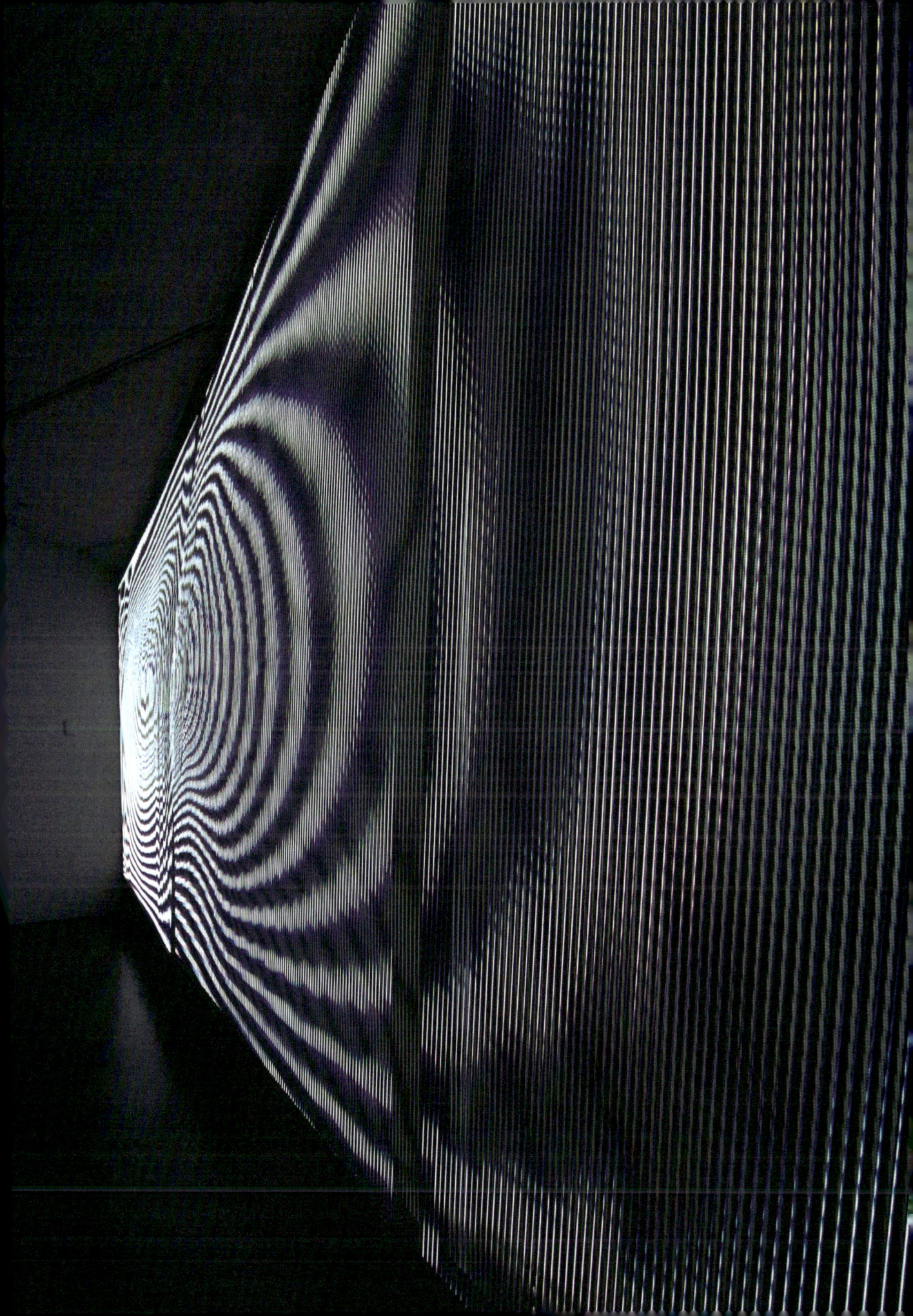

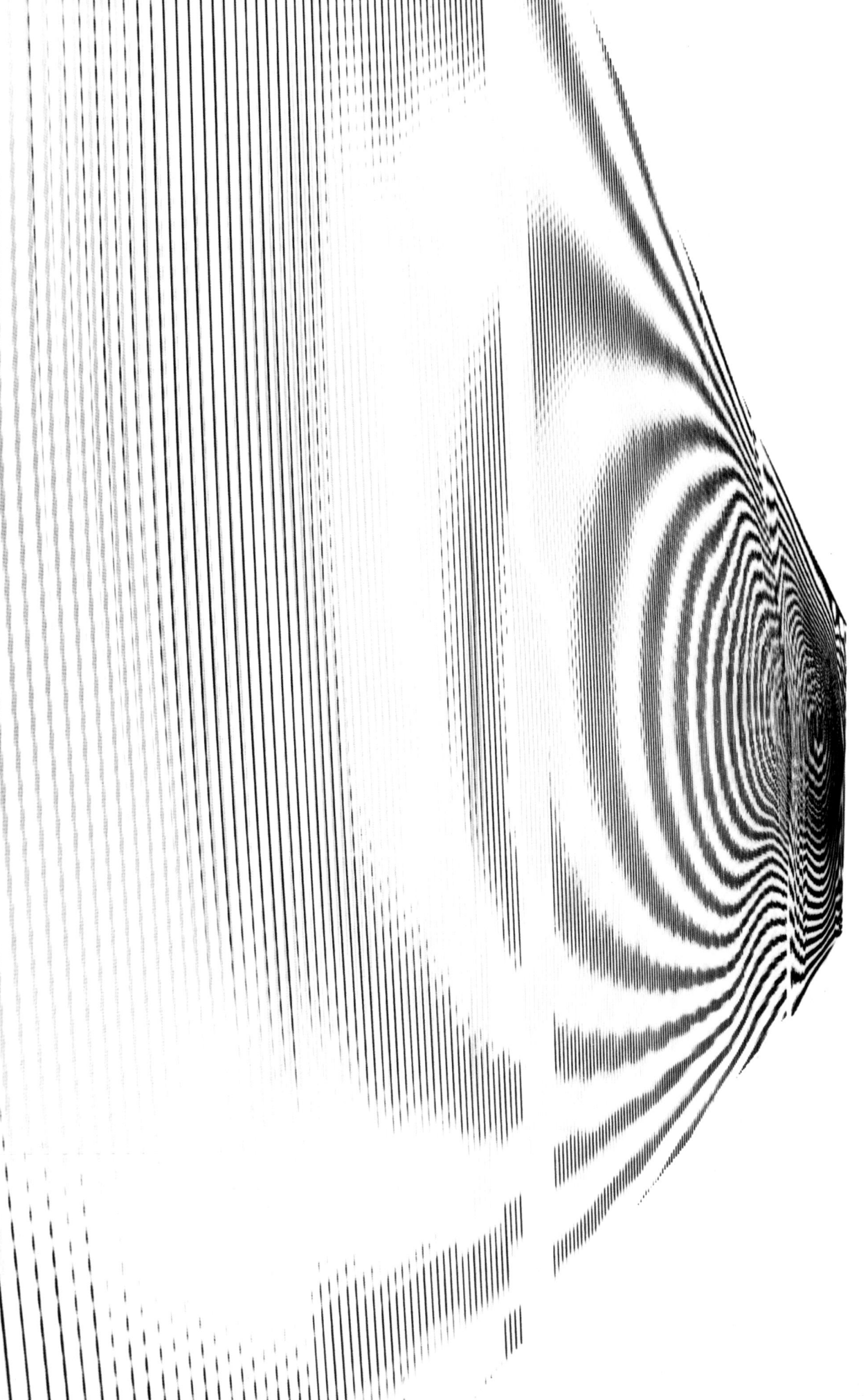

John Pendry, theoretical physicist

LIGHT IS A DANCE

CLOAKING   HARRY POTTER   DANCE   INVISIBILITY   BENDING LIGHT

ASTRID ALBEN   Good evening listeners, my name is
Astrid Alben from PARS and I'm here in the studio with
Sir John Pendry, the eminent physicist known to the public
primarily for his theoretical work on the characteristics
and behaviour of light, and for making the visible *in*visible
with metamaterials, like the cloaking most of us will know
from the Harry Potter films. Welcome John.

JOHN PENDRY   Hello!

AA   How would you explain invisibility to a non-physicist?

JP   Invisibility involves two challenges. One is that the
object you make invisible shouldn't reflect any light itself so
you can't see it. That's relatively easy. A pot of black paint
will do that job. And that's how conventional stealth works
with modern aircraft; the radar doesn't bounce off them – it
doesn't bounce off it in a direction that will be detected.

But there's a more serious challenge, which is that you
shouldn't be able to detect that the thing is being hidden,
even if you can't see it; you want to be unaware that it is
being hidden and the way you detect something that's black
is through a shadow. And the challenge is to get rid of the
shadow. It's the Peter Pan question if you like. You've got
to lose the shadow.

AA   Okay, so you have to lose the shadow. So the object
can't reflect itself on to another surface?

JP   Not quite ... it can't obstruct light that wants to get into
your eye. So if I want to hide you – any light that's coming
from a source behind your body – I've got to take that light,
guide it around you, and then send it on to the course that
it would take if you hadn't been there, and that's quite a
challenge. Because it has to remember where it comes from.

AA Professor Pendry, if I understand correctly, there are materials that are called metamaterials that have the ability to not reflect light but they somehow bend it?

JP They can do that, but the point about metamaterials is that they can do almost anything because they extend the properties which are available in ordinary materials like glass, which interact with light but not in the way we want. And metamaterials are a new class of materials that enable properties.

So the first challenge is to design the materials that you want to do the cloaking procedure with and then you've got to find a way of making them because they're not there in nature. You've got to use this metamaterial procedure to construct the cloak.

So there are two stages: designing the cloak and then building it. And to build it you need a material, and that material happens to be called a metamaterial.

AA But the metamaterial does not yet exist.

JP It does exist because we've built a cloak.

AA Okay so it does exist.

JP Yes it does exist but not in nature.

AA So there are no animals or plants, no organic matter that has this quality.

JP I know of no invisible animal. [*He laughs*]

AA Okay, just checking.

JP Yes. [*Both laugh*]

AA And if there's a question that you would like to have answered about the world, what would that question be?

JP I think the greatest challenge of the moment in science is, what is the nature of information? And, as an artist you will have your own definition of what information is. And be very clear about it: but as a scientist you have to think about real quantities, like information as physical objects. I compare it with things like energy, for example; we know what energy is and we know that it has various forms.

We know something about information but we don't have many physical laws that will describe how it behaves. That is a challenge and it's a challenge that the quantum world I believe is going to answer – I hope in my lifetime – but it's a very difficult problem.

AA And do you think you could formulate that question for me?

JP No. If I could I think I could answer it. My wife – who is also a scientist – said at one stage that if you can formulate the question you've usually answered the question at the same time. So it's a very, very difficult problem. And we don't even know the right question to ask at the moment.

AA Could it be that it's the same for the solution? I think it was Edward Witten, the scientist who united String Theory; he said that once he had the solution to the problem of

String Theory that it was so simple he didn't need to write it down. This sense of elegance, well, he called it simplicity, is it important in science? It is in art and our appreciation of beauty. What does elegance mean to you in your work?

JP Yes, there's a belief amongst physicists that I suppose Paul Dirac articulated well. He said it's more important that your equations are more beautiful than that they agree with the experiment. I think that's a bit extreme myself. But it is often the case that when you're wrestling with a difficult problem there are various answers you try, and the one that is correct very often has a certain beauty to it. By beauty we mean such attributes as symmetry, simplicity, but that's not always true. I think that there are particularly mathematical problems that have defied solution for a long time simply because there isn't an elegant solution and we've been looking for the elegant solution and it just isn't there.

In my case, I think the simplest thing I did which has had the most impact on my colleagues is to realize this: In optics we have the principle that if you're trying to see something very small with an ordinary microscope, you can't see anything smaller than the wavelength of light to the first approximation. And that's a huge restriction because the contents of a living cell are all smaller than that limitation and that is a very, very severe limitation. People believe that it is a law of science that it is so. But about fifteen years ago, I was thinking about some of the consequences of my metamaterials, particularly a strange attribute of negative refraction (they bend light the wrong way, sometimes – not always) and I discovered that you could design a lens that didn't obey this limitation. That if you could build it – and build it perfectly – it would focus light perfectly and you could see anything with it!

AA What's the catch?

JP Of course the catch is: if you could build it perfectly. But the realization was you could do that with mathematics that any of my third-year students could have done. It is the sort of problem that has the elegance that once you've seen it, you think: why didn't I think of that before? And people who heard my solution either said, "It's got to be wrong!" [laughs] because it's so simple and obvious, or they said, "Gosh, why didn't we all think of it before?" And that is something which I found very beautiful and has a lot of impact as the consequence of that.

AA Yes, I think things that express themselves in an elegant way or that can be expressed in an elegant way communicate better.

JP Ah, yes, you've put your finger on it there because simplicity also implies that it is communicable in a simple form and therefore a straightforward way and therefore your ideas will reach further. I've done work in the past that has been much more complex than that but has been [chuckles] widely ignored [still chuckling]. Even though I think it's beautiful. It's just too complex for everybody to be bothered to understand it.

AA And if you were to wake up 20 years from now, and your work has found an application in our day-to-day lives – well not necessarily in our day-to-day lives but has found an application – what would that be? How will we be using metamaterials?

JP I think if I could tell you the answer to that now, the applications would probably be trivial because they'd be easy to think of. What I hope is that if I woke up 20 years from now, there would be applications that I wouldsay, "Wow, why didn't I think of that?" Which would surprise me so much and I think there will be such applications because – we alluded to the beauty of an idea and I mentioned two things. One was simplicity and the other was symmetry. But there's a third attribute and that is if you look at the idea and start thinking about it, it has to have a far horizon. So even though it's simple and you think your way through it, you can't see an end to where your thoughts might go. And so, as well as simplicity, there's an inherent, let me say, I was going to say complexity but that's a contradiction, isn't it?

AA Continuance?

JP Potential, let's say. And I think these ideas which we've been working on recently do have potential for others to pick them up and do things with them that we never envisage. I always try to make an analogy with the development of an idea with the development of the laser. And that's a very nice example because it started off as realizing a very fundamental postulate of Einstein's.

And when the thing got going people were amazed by it and they thought, "that's beautiful!" you know, and here's an equation and here's the kit that does its stuff. But they weren't convinced that it was useful for anything. It was called "Money Acquisition Scheme for Expensive Research" = MASER, in its first instance [*laughs*].

But, look what a laser does today: it's the basis of the Internet, it does welds in cars, it'll do eye surgery, there's just no end to the things it could do. And the inventors certainly didn't have those in mind when they invented it but they knew that it was so radical that it had potential.

AA A researcher at CERN told me that they've tried to figure out or calculate how long it takes for a scientific discovery to find an application, if it indeed needs an application.

JP What was their answer? How many years?

AA Seventy years.

JP Seventy?! Oh wow! Yes, well, it can be shorter than that but yes, it's a long time.

AA Where does your fascination with light stem from?

JP Well I've always been fascinated by light. Not necessarily in the scientific sense. I think everybody has some sort of relationship with light. Because it affects our moods so much as well as gives us information. The bald statement that you read using light, you see people, but also just shapes and particularly colours are influencing particularly emotionally, and that is my first recollection of light, and so a little while ago I was giving an interview and a lady asked me a similar question: "What was your first memory of light?" And it was lying in my bed as a child in the morning and our room, my bedroom, faced north. But opposite the bedroom window a little way away was a tall, brick building with lovely orange bricks. And I knew that if I saw this gorgeous orange glow in the morning that the sun was shining and it was going to be a great day. And happiness would come forth and so on and so forth. So that is my early recollection of light. And to this day I love to experiment with photography and particularly with the play of light and how it changes the way you see a scene.

I was walking through London this morning from my flat and the light was absolutely gorgeous, particularly the early-morning light and I whipped my camera out about five or six times and I took some shots with it.

AA   Your story reminds me of a dilemma I'm trying to solve and maybe you can help me: I would love to catch moonlight in a box but so far it's not working. I went and asked a carpenter, hoping he'd be able to build me a strong enough box and he said that this was not possible. So then I went and spoke to an engineer to see if maybe he could help me build a box made of a material that would trap the light. I went to a scientist who works with solar energy. Maybe she could help me harness moonlight. I spoke with a biochemist. Even an astronaut, to bring me back some moonlight from space, and none of these – everyone said "You can't put moonlight in a box".

JP   Then?

AA   Well now I'm speaking to you. [*He laughs*] And you can *bend* light. So maybe you can or at least you can maybe help me with these metamaterials to somehow – can you, could you help me get moonlight in a box?

JP   Firstly, I want to know what you want to do with it once you have it in a box? Because the nature of the box will be determined by – what will you do with your moonlight?

AA   If I have moonlight in the box, I would like to invite people to gather round – on a very dark night. And in an open-air theatre, a Greek theatre, and I would have the box stand in the middle of the stage and I would release the moonlight.

JP   Release it. Yes, yes, wonderful. Release it, okay.

AA   Can you imagine the beam of moonlight being released back to the stars?

JP   Yes, yes, that's ahhh, well, you know light travels so quick. It's very hard to store it. You can capture things which are massive but light has no mass so it will not be tamed, I'm afraid. So all those people you spoke to, um, they were right. So, only in your dreams, as they say.

AA And for now. So when you said that: light cannot be stored because it is so fast; is light, for you, a form of information?

JP It can convey information and it does so in volumes through the Internet. And you could argue that light, which is coded into on-and-off states, if you like, is the purest form of information. And when I worked on information myself a long time ago – that was my model: to say, "If I put this information on a beam of light," and then ask, "what are the physical properties of that light?" So capturing the information in the light and then asking what the light does was my way of analyzing what are the physical properties of the information. How does that information change the properties of the light? How does light with information encoded in it differ from light which is just a beam which tells you nothing? Through that analysis you could say certain things about how information moved – was there any energy associated with it and did you have to pay an energy price to shift information around? Questions like that, very simple questions, but they're very, very fundamental.

There's another question: you asked me, earlier on, can I capture moonlight in a box? Well, you can ask the question, can you capture information in a box? You can say that you can keep track of energy because you can say, if this volume of space loses some energy, you can track where it's going through the flow of energy. You can define a flow of energy. But you can't really do the same thing with information. Because information doesn't sit in one place, it's what we call non-local. It's everywhere and nowhere if you like. And that's one of the paradoxes of quantum mechanics. Quantum mechanics is not a local theory. And the answer to "What is information?" lies somewhere inside quantum mechanics in a way that certainly I don't understand at the moment and I don't think anybody does.

AA So the thing that feeds human curiosity, which is information, is also very elusive, slippery and certainly you can't put it in a box and you can't put moonlight in a box.

JP [*Gentle chuckle*]

AA That's a pity ... but you know, poets have a way of avoiding the restrictions of reality. [*They both laugh*] One final question. Professor Pendry: do you have a simple definition of light?

JP Yes: it's a dance. It's a dance between two entities: one is magnetism and the other is electricity. Just over 150 years ago, James Clark Maxwell showed what light was and until that time, people used light, they measured its speed, all those sorts of things. But they did not know what it was. And what he found out through his equations, through studying electricity and magnetism, was that if you twine a magnetic field around an electric field you can throw it like a bubble in space and it propagates forever. And I think that's the most wonderful revelation – for me, being interested in light – is the most wonderful revelation that physics has produced.

Imagine his excitement and he didn't know it was true until he did an experiment. He was not a theorist, he did experiments and he knew that his equations that he had for electricity and magnetism made a wave but he didn't know that the wave, was light – suspected it might be but he didn't know, and he needed to do an experiment which measured the speed of the waves which he'd found. And it was an electrical experiment; he had to measure a quantity to do with the capacitance of a big brass object, actually. And when he got that number he found that it agreed within a few percent with a known velocity of light so bang on he knew that was it. And talk about Eureka moments! You know, thousands of years of studying this stuff and we didn't know what it was. And then, he found out. It's wonderful.

AA Thank you very much, Professor Pendry.

JP Thank you!

Dinie Besems, jewellery designer

A SLICE OF WATER

GLISTENING   TACTILE   RANDOM

If only a plus and a minus and one battery are necessary for light, then you can have many, many LEDs linked together with silver wire, as a flexible pouch that lights up when you hold it. I connected only the plus. Your hands will be the minus. As soon as the LEDs connect with the minus they light up. Always at random. Never to serve as a light in the dark but to create a poetic moment in your hand, like a slice of gorgeous, glistening water.

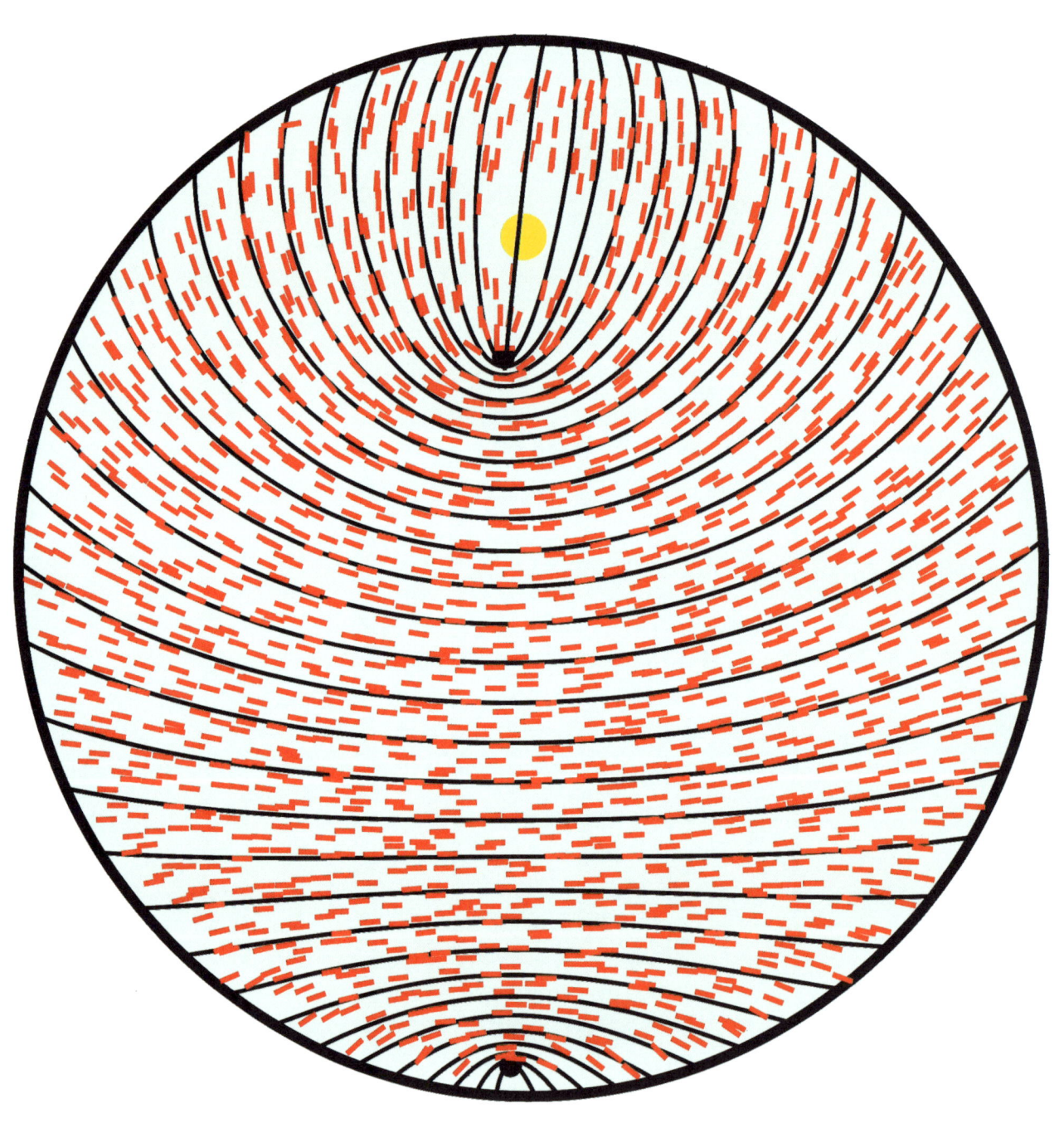

Michael Berry, physicist

THE ELLIPTIC INTEGRAL
IN THE SKY

POLARIZATION FINGERPRINTS PATTERNS SKY SCATTERING
YELLOW SUN NAVIGATION

The light of the day sky is sunshine scattered by air. It is blue because the blue waves in the spectrum from the sun are shorter than the red waves, so an air molecule appears bigger to arriving blue waves and scatters them more strongly. But there is more to daylight than the brightness and colour that we see. There is a hidden property – *polarization* – that we can scarcely perceive (though some creatures – for example bees – can). Polarization is the sideways vibration of the electric and magnetic fields comprising light. In the sky is a secret pattern of polarization, built from four fingerprint structures.

The starting-point for understanding this is that light arriving from the sun is a random assembly of transverse vibrations in all directions – not polarized at all. But the act of scattering induces polarization. Each molecule is a tiny aerial, set into vibration by incoming light and re-radiating it in all directions. Different incoming polarizations are scattered differently, in a way that appears strongest at right angles to the sun, and weakest in the direction of the sun and opposite to the sun (the "anti-sun", visible for a while before sunrise and after sunset). This can be seen by looking at the sky through a polarizing sheet, such as a lens from polaroid sunglasses, and rotating it. The brightness varies, more powerfully at 90° from the sun and hardly at all near the sun and the anti-sun.

single scattering

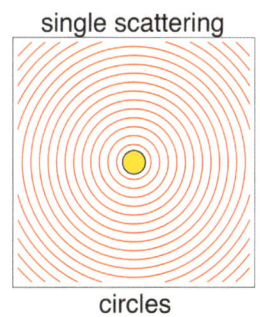

01   circles

polarization
⫸ ⟶
splitting

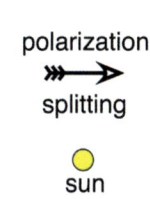
sun

multiple scattering

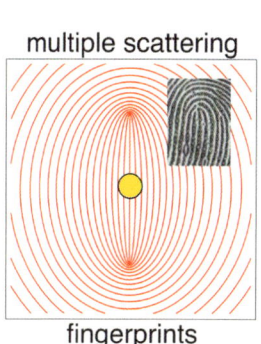

fingerprints

In this theory, based on the sunlight being scattered once, the polarization pattern of daylight would be structured around the two unpolarized antipodal points in the sky: the sun and the anti-sun. But the reality is different. Already in the early 19th century – and long before the scattering of polarized light was understood – three unpolarized points were observed in the sky: above and below the sun, and above the anti-sun. A fourth was predicted, below the anti-sun, and it was observed in 2002, from a balloon flying after sunset. The reason for four unpolarized points, rather than two, is that light is scattered more than once. This multiple scattering is weak, but its effect is to split each of the unpolarized points in the one-scattering theory into two.

The four unpolarized points, and the pattern of polarization across the sky, emerged from elaborate multiple-scattering calculations. But these features can be understood in a simpler way, based on considering the geometric nature of the unpolarized points. Around each unpolarized sky point in the one-scattering theory, the direction of polarization (for example of the electric field of the light) turns once, and in the same sense. In mathematical terminology, such an unpolarized point is a *singularity with index +1*. When multiple scattering splits it into two, the resulting unpolarized points are singularities of index +½: around each of them, the polarization makes a half-turn. This gives rise to a local pattern like the ridges on a fingerprint, and the reason is similar: polarization is not a direction with an arrow (that is, not a vector), but rather an oriented line – direction without a sense – which looks the same after a half-turn as well as a full turn. That is why index +½ is natural for polarization. [01]

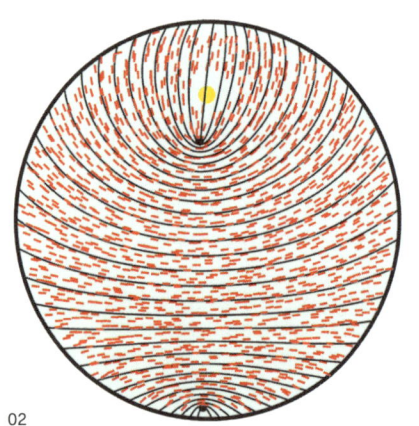

02

The full polarization pattern lives on the full sky sphere of directions seen from Earth, and is based on the four finger-prints. Of course we don't see the full sphere: at any one time, our sky is the hemisphere above us, with just two of the fingerprints. On the full sphere, the pattern can be built by interpolating between the two antipodal fingerprint pairs of unpolarized points. The simplest way of doing this was devised in 2004 in collaboration with Mark Dennis, and agrees accurately with measurements of the sky by Raymond Lee at Annapolis, Maryland on 3 November 2003. [02]

The yellow sun was 33.7° above the horizon. The measured polarization directions are the little red line segments. The pattern predicted from theory is represented by the black contour lines in the sky: at each sky point, the polarization direction is parallel to the contour passing through that point. Contour lines of what? Of a function built from *elliptic integrals*. As the name implies, these were originally devised in the 19th century to calculate the length of the perimeters of ellipses. But they have many other applications, for example to the bending of elastic wires, light rays refracted by sound waves, the statistical mechanics of particles on lattices ... Their appearance in the sky was unexpected.

I wish this nice picture was the one we published. It wasn't. Instead, in the version we published, the observed polarizations – the little red line segments – were plotted as Raymond Lee measured them: on a rectangular grid in the sky. But the grid tends to deceive the eye: we see the lines of the grid of line segments, rather than the orientations of the individual segments. The agreement between theory and experiment is much less clear. An obvious fix is to randomize the positions of the line segments [03] within each pixel, while, of course, preserving their orientations, namely the polarization directions. This rendering by Mark Dennis eliminates the distracting effect of the grid, while preserving the data to the accuracy with which it was measured.

$$\text{contours:} \quad \operatorname{Im} F\left[\arcsin\left[i\,\frac{x + i\,(y + y_s)}{A\,(1 + iy_s\,(x + i\,y))}\right], \frac{1}{A^4}\right] = \text{constant}$$

F = elliptic integral of the first kind, x, y = coordinates in the sky, A = fingerprint splitting, y_s = sun elevation

We cannot see this pattern directly, but with polarizing crystals we can detect the polarization direction at each point in the sky. And although the basic theory was devised by considering the clear-blue sky, it is remarkable that the pattern of polarization directions persists even in cloudy or foggy skies, although the strength of the polarization is weakened. This has reinforced the speculation that the Vikings might have used polarization to determine the position of the sun, aiding

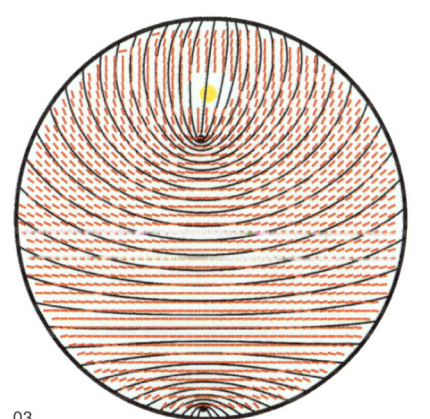
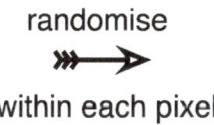
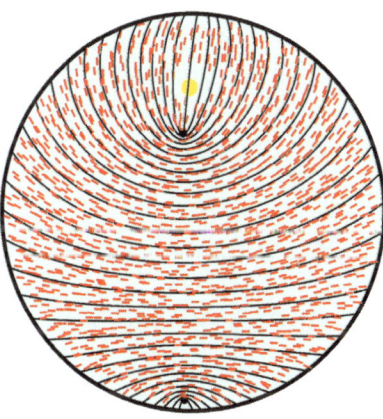

randomise
within each pixel

03

navigation westward from Norway via Iceland to what is now Canada. I share the scepticism of historians about this conjecture, because the only historical evidence that might support it is an enigmatic mention of a "sunstone". Nevertheless, it is interesting to explore "what if" history, and investigate ways in which polarization could conceivably have been used, for example using Iceland spar, polarizing crystals that can be found naturally.

1   M.V. Berry, M.R. Dennis, R.L.J. Lee (2004), "Polarization singularities in the clear sky", *New J of Phys*, 6:162
2   G. Horvath, B. Bernath, B. Suhai, A. Barta, (2002), "First observation of the fourth neutral polarization point in the atmosphere", *J Opt Soc Amer A*, 19:2085-2099
3   R. Hegedus, S. Akesson, R. Wehner, G. Horvath (2007), "Could Vikings have navigated under foggy and cloudy conditions by skylight polarization? On the atmospheric optical prerequisites of polarimetric Viking navigation under foggy and cloudy skies", *Proc R Soc A*, 463:1081-1095
4   L.K. Karlsen (2003), *Secrets of the Viking Navigators*, One Earth Press, Seattle

Guy Ropars, Albert le Floch, physicistst

A LATE DISCOVERY OF AN EARLY VIKING NAVIGATION TOOL

SUNSTONE VIKINGS NAVIGATION ICELAND SPAR DISCOVERY
POLARIZATION TOOL

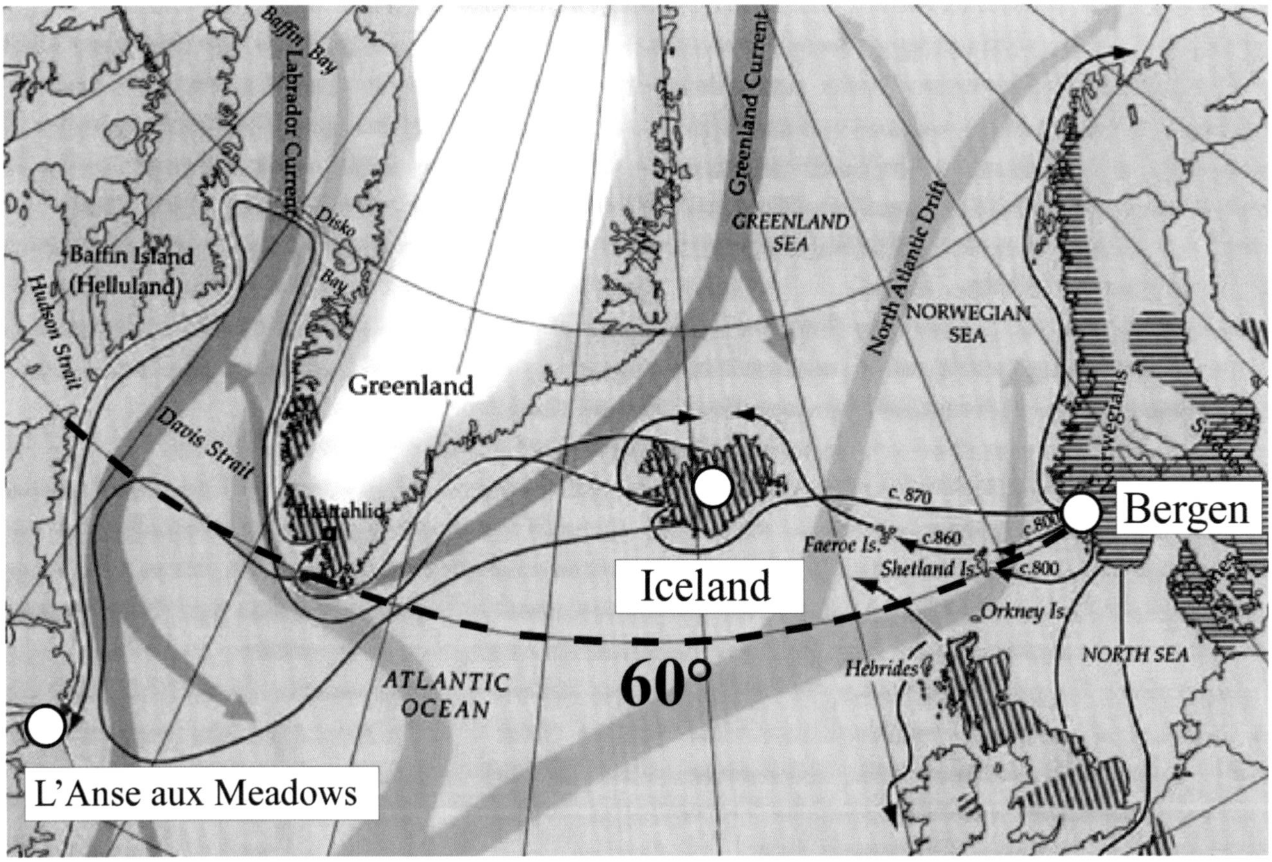

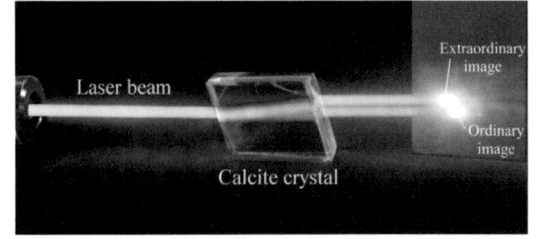

01–02

It is well known that Vikings reached Newfoundland in America around 1000 CE.[1] One may wonder how the Vikings successfully navigated around the 60°N latitude, crossing thousands of kilometers of open sea [01], especially since the magnetic compass was only introduced in Europe in the 13th century. Of course, Vikings could have navigated using the sun (sundials), the stars, the direction of the wind, the waves and the swell. But at around 60°N, in summer the stars are only visible for a few hours. Moreover, the sun can be hidden behind the clouds or can be below the horizon. Even so, the Icelandic Sagas of the 13th century tell us that Vikings were able to locate the sun azimuth by looking at the zenith through an enigmatic *sunstone*.

THE POLARIZATION OF THE LIGHT

In contrast to the *longitudinal* nature of sound waves, the *transverse* nature of light waves (waves on a string) introduced the concept of the polarization of light, ie, the axis of vibration of an associated electric field in the 19th century. However, curiously, without any knowledge of this fundamental property of light, many animals use the polarization every day for their orientation, as Vikings might have used it in their navigation from Scandinavia to America five centuries before Columbus.

Polarization is the third fundamental property of a light wave, after intensity and frequency (which allows us to perceive colour). In 1808, observing the light, reflected from a window of the Palais du Luxembourg in Paris through an Iceland spar (calcite crystal), French officer, engineer and polymath Étienne-Louis Malus realized that this light, reflected by a glass plate, was different from the usual natural light. Malus introduced for the first time the word *polarization* to characterize this newly discovered property of light, which is more complex than intensity and frequency. A year later, in 1809, François Arago discovered that skylight at the zenith was also polarized. Today polarized light is ubiquitous both in nature optics[2] and in many optical systems. Most of our computer screens are polarized at –45° from the vertical. Moreover, the polarization of electromagnetic waves is used for instance in lasers, matter-light interactions and astrophysics.

POLARIZATION OF THE SKYLIGHT
AND THE VIKING SUNSTONE

Drawing an analogy with a polaroid-based instrument called the Twilight Compass used by Scandinavian Air Systems trans-Artic pilots, the Danish archaeologist T. Ramskou suggested in 1967 that Vikings might have used polarization for their navigation.[3] He proposed that a mineral common in Scandinavia, like cordierite, tourmaline or calcite crystals, could have been used as sunstone. Unfortunately, he supposed that all these crystals behaved as absorbing polarizers, which transmit light vibrating along one direction while absorbing the light along the orthogonal direction, like those used in Scandinavian aircraft. This is true for cordierite and tourmaline, but their absorption limits the sensitivity of the observations. By contrast, calcite is completely transparent without any absorption of the light.

In 1669, the scientist and physician Erasmus Bartholinus discovered the *double refraction* of calcite. Calcite renders two images at the output, the ordinary and the extraordinary image [02]. Unfortunately, because polarization was not discovered until 1808, physicists like Bartholinus, Newton and Huygens merely wrote that light has *two opposite sides*. By simply rotating a calcite crystal in front of a window of the Palais du Luxembourg, Malus had noticed the relative variations of the intensities of the ordinary and extraordinary images and discovered the *polarization of light*. The Vikings may also have been able to rotate an Iceland spar when looking at the zenith and may have observed the same intensity variations so as to deduce the sun's position. Today we know, thanks to the Rayleigh scattering of sunlight through the Earth's atmosphere (which causes the blue hue of the daytime sky and the reddening of the sun at sunset), that, looking at the zenith, if we detect the direction of the polarization of the skylight in sunlight, the sun azimuth is found at exactly 90° from the direction of skylight polarization.

We have shown that looking at the zenith through a calcite crystal gives a powerful differential method to pinpoint polarization.[4] Taking into account the sensitivity of the human eye to contrast, an enhancement of a factor of 100 can be reached by comparison to the method using modern-day absorbing polarizers. Figure [03] shows typical experimental and theoretical curves corresponding to two

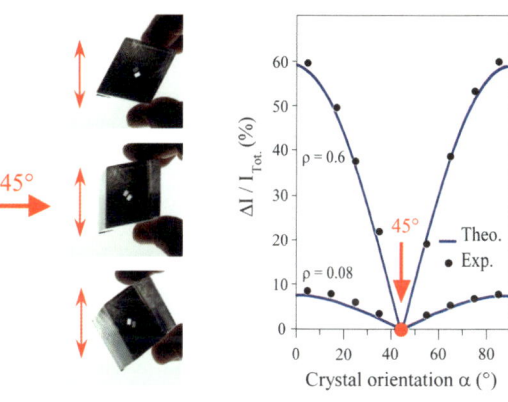

03–05

01 The main Viking sailing routes around the 60°N parallel.
02 Photo of the ordinary and extraordinary images obtained with a blue laser passing through a calcite crystal.
03–05 Experimental differential intensities of the two images at the calcite output, on both sides of the *isotropy point*, ie, α = 45° (top and bottom [04]).

sky conditions encountered in navigation (degree of polarization r = 0.6 corresponding to a clear sky and r = 0.08 corresponding to a rather cloudy sky). The V-shaped curves linked to the *contrast variation*s are particularly well suited for sky polarization pointing.

To test our theory of the Viking calcite sunstone, we built a prototype of a Viking sunstone compass [04]. As shown in [03], one has to rotate the calcite crystal to obtain the same intensity for the two images rendered by the calcite crystal. The arrow in [04], fixed according to a preliminary calibration, gives the direction of the sun. Taking into account the time and the sun azimuth shift of around 15°/hour at the 60°N latitude, the red arrow determines north. Typical experimental sun tracking is reported in [05]. This correlates perfectly with the theoretical values given by the astronomical tables. Please note that the Viking sunstone compass works correctly until the appearance of the first stars in the evening sky. Could Vikings have used the polarization of the light without any knowledge of the polarization concept? Surprisingly, looking at the animal world, we know that bees, spiders and beetles, thanks to eyes dedicated to polarization detection, also use the polarization of the sky for their navigation. Moreover, some animals, like beetles, are even able to use the weak polarized moonlight scattered by the Earth's atmosphere.[5]

As a final question, one may wonder if such a Viking calcite sunstone has ever been recovered from a Viking settlement. During a recent excavation in Iceland, a small calcite fragment was indeed discovered in a Viking settlement, giving evidence that some people in the Viking age used Iceland spar crystals. Furthermore, a large calcite crystal was recovered from a 16th century shipwreck.[6] To the best of our knowledge, it is the sole calcite crystal found on board of an ancient ship. Although this ship went down off the coast of Alderney (English Channel) five centuries after the Viking age, this crystal could have been used in navigation, so as to correct possible spurious magnetic deviations, the calcite crystal giving an absolute reference. Moreover, the Alderney crystal brings additional interesting indications for searching sunstones in Viking shipwrecks. Indeed, the so-called *ionic exchanges* occurring in seawater reduce the transparency of a calcite crystal but reinforce its geometry with its specific 102° and 78° angles.

Curiously, the high sensitivity of the Iceland spar to polarization, with its ordinary and extraordinary beams, is currently used to detect the atmospheres of exoplanets, in spite of the extremely weak intensities of light reaching the Nordic Optical Telescope at La Palma (Canary Islands),[7] the existence of an atmosphere being a condition for possible extraterrestrial life.

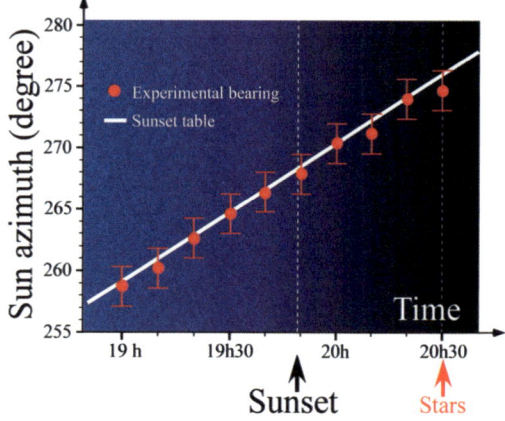

04–05

1   R. Nydal (1989), "A critical review of radiocarbon dating of a Norse settlement at l'Anse aux Meadows, Newfoundland Canada", *Radiocarbon*, 31 (3):976–985
2   M.V. Berry (2015), "Nature optics and our understanding of light", *Contemporary Physics*, 56:2–16
3   T. Ramskou (1969), *Solstenen. Primitiv navigation I Norden før kompasset*, Rhodos, Copenhagen, Denmark (An English summary is included)
4   G. Ropars, V. Lakshminarayanan, A. Le Floch (2014), "The sunstone and polarised skylight: ancient Viking navigational tools?", *Contemporary Physics*, 55, 30:302–317
5   M. Dacke, D.-E. Nilsson, C.H. Scholtz, M. Byrne, E.J. Warrant (2003), "Animal behaviour: insect orientation to polarized moonlight", *Nature*, 424:33
6   A. Le Floch, G. Ropars, J. Lucas, S. Wright, T. Davenport, M. Corfield, M. Harrisson (2013), "The sixteenth century Alderney crystal: a calcite as an efficient reference optical compass?", *Proceedings of the Royal Society A*, 469 20120651
7   S.V. Berdyugina, A.V. Berdyugin, D.M. Fluri, V. Piirola (2011), "Polarized reflected light from the exoplanet HD189733b: first multicolour observations and confirmation of detection", *The Astrophysical Journal Letters*, 728:L6

04   Prototype of a Viking sunstone compass.
05   Typical experimental sun tracking (below the horizon).

Tamara Frank, marine biologist

DON'T LOOK AT ME!

DEEPSEA PHOTOPHORES BIOLUMINESCENCE VISUAL PIGMENT
PREDATOR SEDUCTION

Deepsea organisms have developed a number of ways to see light. But the light they see might not be the light that you and I are used to seeing. Bioluminescence is relatively rare on land, being found only in fireflies, glowworms (larval insects), and a few species of beetles. However, it has been estimated that 80–90% of the animals living in the ocean at depths between 200 and 1000 m are bioluminescent, so many of the visual adaptions found in deepsea organisms may be adaptations for seeing bioluminescence rather than downwelling light from the sun. As you go deeper, seeing the contrast of a body against the dim remaining light from the sun becomes more and more difficult, while seeing bright point sources of light from bioluminescent animals and seeing dim glows from mats of decaying dead stuff colonized by bioluminescent bacteria becomes easier and easier.

One way to see light is to have visual pigments designed for maximum sensitivity to the light in your environment. Humans, which have access to all colours of the rainbow, from 400 nm (violet) to 700 nm (near-infrared), have three cones of visual pigments, absorbing maximally at short wavelengths (blues), medium wavelengths (greens) and long wavelengths (reds). Further integration of signals coming from these different photoreceptors gives us our remarkable colour sensitivity. Having multiple visual pigments is common in shallow-water marine species as well, as they have full access to visible light coming from the sun. However, as you go deeper into the water column, wavelength-dependent absorption and scattering occurs. In the clearest ocean waters, blue light at around 480 nm is attenuated the least, so light of this wavelength penetrates the best. In addition, most of the bioluminescence also falls in the blue wavelengths. Therefore, for optimum sensitivity to the available light from both downwelling light and bioluminescence, one would expect deepsea animals to have replaced all those visual pigments that are sensitive to wavelengths (colours) that are no longer available with a blue-sensitive visual pigment.

01–02

The eyes of deepsea species are not easy to work on, because they are extremely sensitive to light, so special collecting techniques must be used to bring them up from the depths without blinding them. I have a Tucker Trawl [01] with a temperature-insulated cod-end (collecting vessel) [02] at the end that can be closed at depth, ensuring that deepsea animals are brought to the surface in the dark and in cold water. While most people think that it is the pressure difference that kills animals brought up from the depths, this really isn't a problem for animals living as deep at 1500 m, as long as they don't have swim bladders. However, temperature changes can be lethal, so it is important that they be brought up dark and cold. My research interest lies with the crustaceans (shrimp, krill, crabs) because they are among the most robust deepsea species.

So how do I study the photosensitivity of deepsea species, since I don't speak crustacean? I use a device called a monochromator [03], which splits white light into its component wavelengths, together with a neutral density wheel under computer control, to shine lights of

01   3 m² opening/closing Tucker Trawl.
02   Temperature insulated, light-tight removable cod-end on the end of the Tucker Trawl.

04–05

different wavelengths (colours) and irradiances on their eyes. Using a microelectrode placed on the outer surface of the eye or just below the cornea, I can record an electrophysiological response called an electroretinogram (ERG), which varies in size depending on how sensitive the eye is to that particular colour. In addition to the light source, my electrophysiological rig consists of a microelectrode

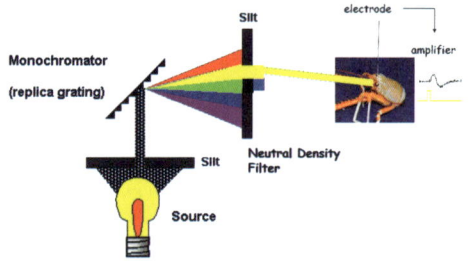

03

amplifier to amplify the microvolt signals, an electromagnetic shutter under computer control to vary the duration of the flash, a Faraday cage to block out extraneous noise that might be picked up by the very sensitive microelectrodes, a refrigerated water bath, so I can keep the animals at their ambient temperatures of 4° to 70°C, and a data acquisition system [04]. It's always best to conduct physiological work on fresh animals, so I bring my electrophysiological rig out to sea with me. Research cruises are usually filled with a number of researchers working on different projects, and since I can't expect them all to work in the dark just because my animals need to be kept dark, I usually end up making a dark room [05] over the Faraday cage, constructed of an ocean-going scientist's favourite accessories – duct tape, black plastic sheeting and PVC pipe. In order to determine the spectral sensitivity of various species, I record a criterion response at 5 nm wavelength intervals from 350 to 650 nm. I adjust the irradiance at each wavelength using the neutral-density wheel until I get a 50 µV criterion at each wavelength tested. The idea behind this process is that, if the species has a blue-sensitive visual pigment, it will take less light at the blue wavelengths to generate the 50 µV criterion than at the violet, yellow, etc wavelengths. I try to get spectral sensitivity curves from at least five individuals of the same species, and in order to combine these data into one average curve, as shown in [06], I normalize the data. This means that, for one individual, I divide all the data by the highest value (maximum sensitivity), so that the values now range from 0 to 1. This gives a much better average curve than trying to combine the raw data. The normalized data are fitted to a visual pigment nomogram, and this then tells me where the peak sensitivity lies. All the species that I have studied to date have the anticipated single blue visual pigment, with a relative spectral

03   Diagram of monochromatic light stimulus.
04   Electrophysiological set-up on shipboard.
05   Electrophysiological set-up in foreground with darkroom built over Faraday cage in background.
06   Average spectral sensitivity curve for a species with a single visual pigment. Bars are standard errors. Solid line is visual pigment nomogram fit to the electrophysiological data, showing a peak sensitivity of 490 nm for this species.
07–09 Oplophorid shrimp with dual visual pigments.
07   *Systellaspis debilis.*
08   *Oplophorus spinosus.*

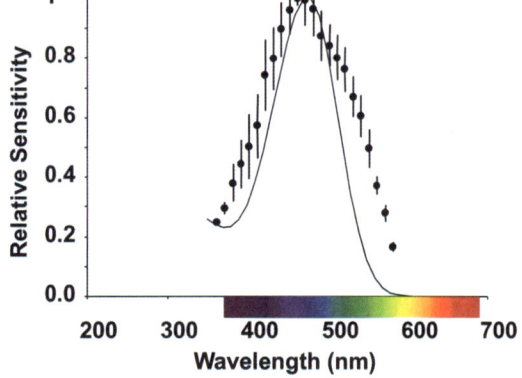

06

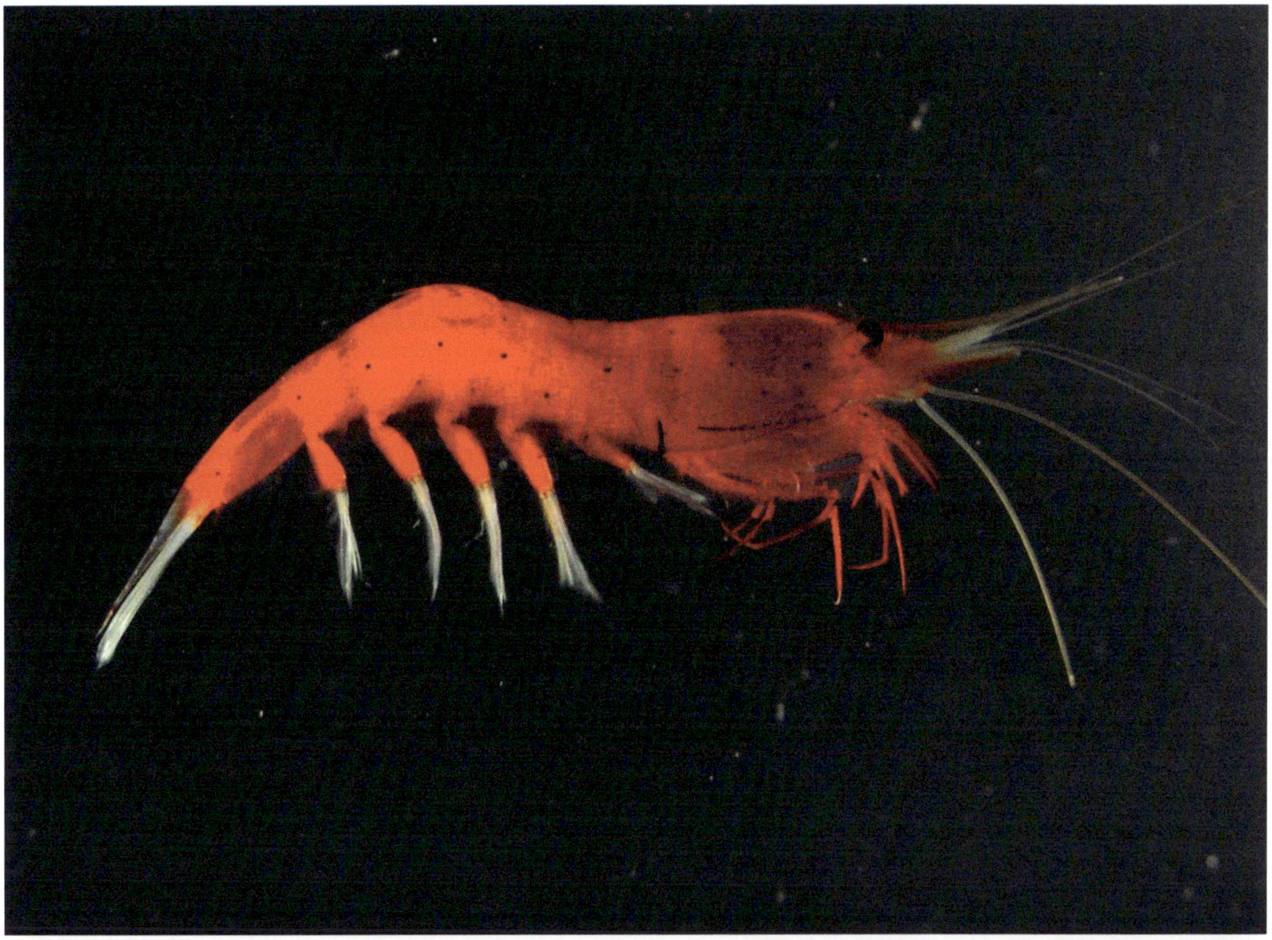

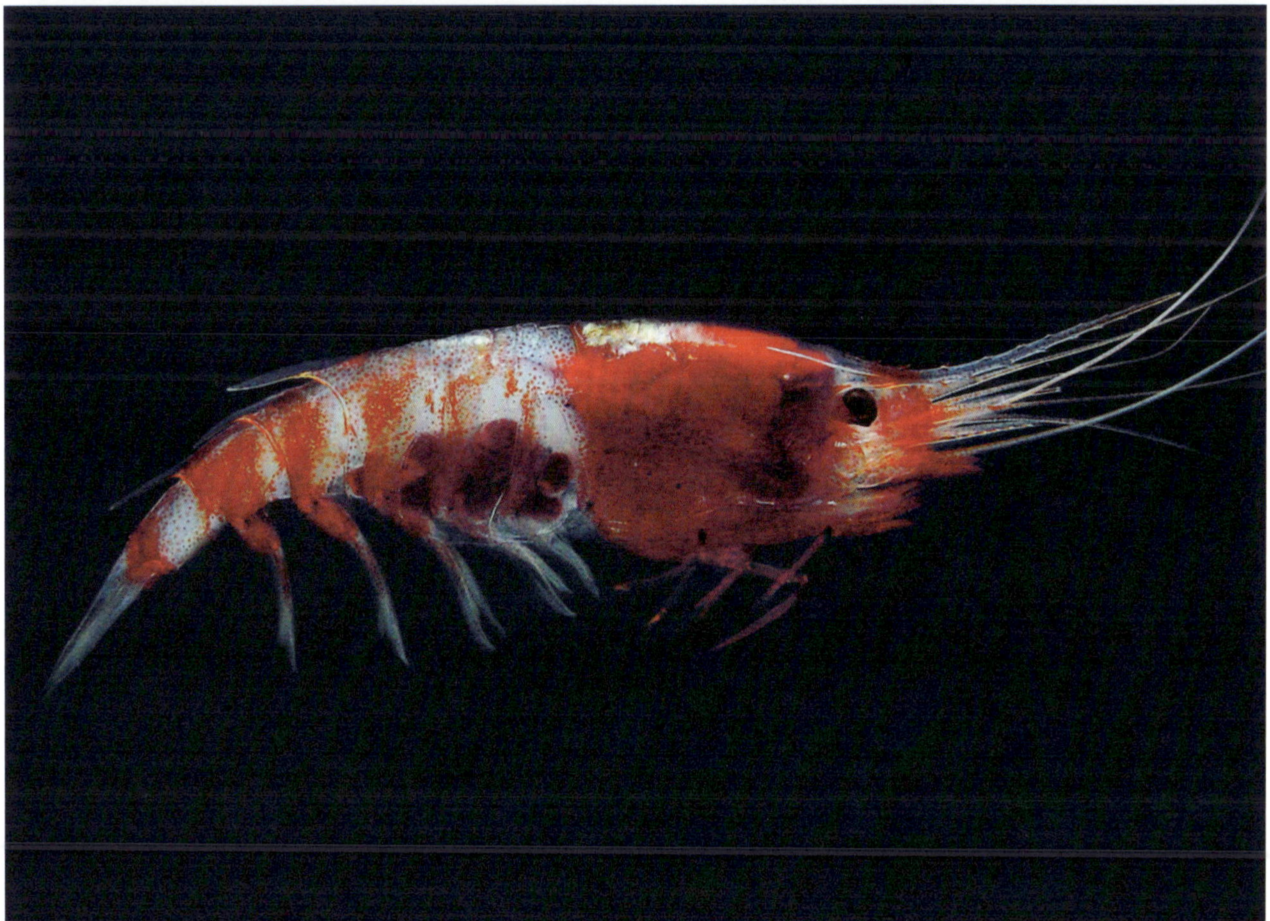

07–08

09–12

sensitivity similar to that shown in the figure [06]. However, there are a few interesting exceptions. There are several species of pelagic (ie, they live in the water column and never touch the bottom) deepsea crustaceans in the family Oplophoridae that have two visual pigments [08-10], one with maximal absorption in the blue, as expected, but the other with maximal absorption in the violet and near-UV wavelengths [13]. These species have daytime depths below 500 m, where any remaining light in the UV wavelengths is far too dim for vision. Interestingly, the other species in the same family [10-12], some with the exact same depth distributions, have single visual pigments.

So, what characteristics do these pelagic deepsea crustaceans with two visual pigments share that set them apart from every other pelagic deepsea crustacean that has been studied so far? The answer has to do with bioluminescence. Everyone is probably familiar with fireflies, which have light-producing organs called photophores. In the oceans there are two forms of bioluminescence – photophores and a bioluminescent spew. The photophores, found on the ventral (belly) surface [14-15], serve several purposes. One verified function is counterillumination. Animals swimming in the water column are easily visible to predators below them, who see their bodies as dark shadows against the downwelling light field. The animals with ventral photophores are able to produce light that exactly matches the light being blocked by their bodies, and their shadows disappear. Another hypothesized function of photophore light is to attract mates, because photophores often occur in species-specific patterns (and this is a well known function of photophores in fireflies). However, this hypothesis has not yet been verified for deepsea species. The other form of bioluminescence is a bioluminescent spew, much like the inky spew that a squid puts out in a tide pool, but this spew is blindingly bright [16]. This serves a purely defensive function, and is found in squid, fish and crustaceans. In shrimp, emission of the spew is always accompanied by a tail flip – the spew momentarily blinds the predator while the shrimp tail flips away in the opposite direction.

A number of crustacean species possess photophores, and a number have a bioluminescent spew, but having both photophores and a spew is extremely rare. The crustaceans with the two visual pigments have both photophores and a spew, and are amongst the only species of deepsea crustaceans known to possess both. Some mathematical modelling by Dr Tom Cronin demonstrated that the two pigments were at the optimal position to tell the difference between spew (Danger! Danger!) and photophore (Come hither!) luminescence, which would be very beneficial to these species.

Among the benthos (animals that live on the bottom), I have also discovered several species with similar unusual adaptations

09   Janicella spinacauda.
10–12 Oplohporid shrimp with single visual pigments.
10   Acanthephyra purpurea.
11   Notostomus gibbosus.
12   Meningodora vesca.
13   Spectral sensitivity curve for a species with two visual pigments. The two peaks of sensitivity are at 400 nm and 490 nm.
14–15 Image of a hatchetfish, Argyropelicus aculeatus, and a species of krill, Euphausia superba, showing arrays of ventral photophores.

13

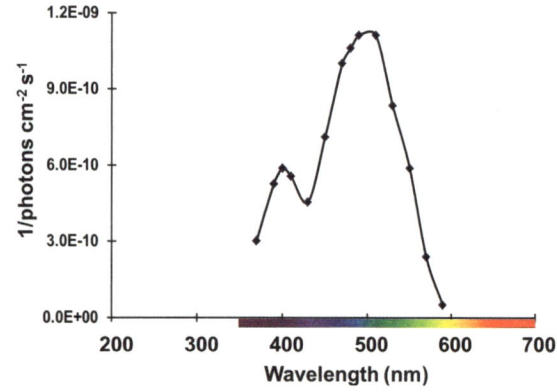

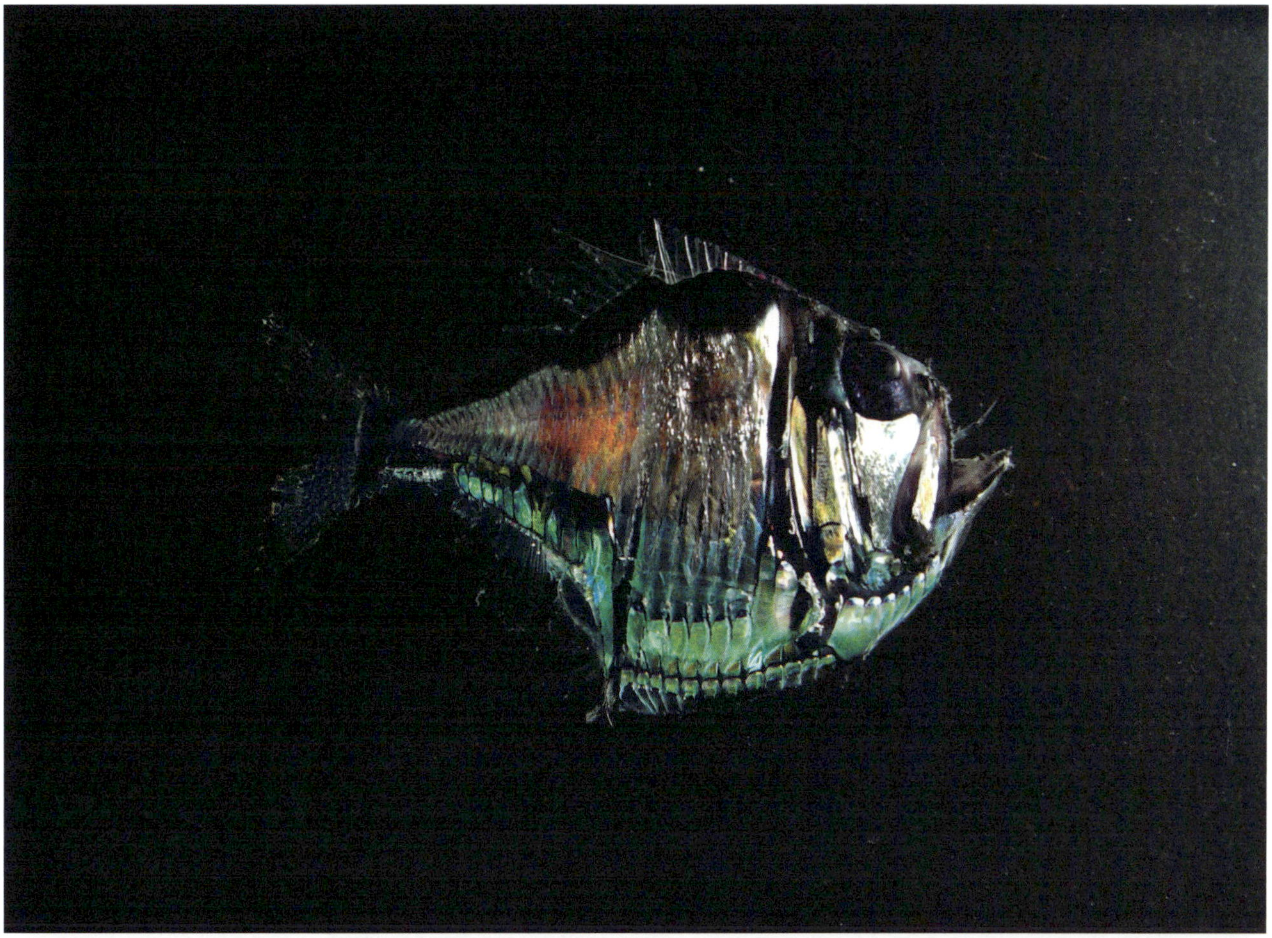

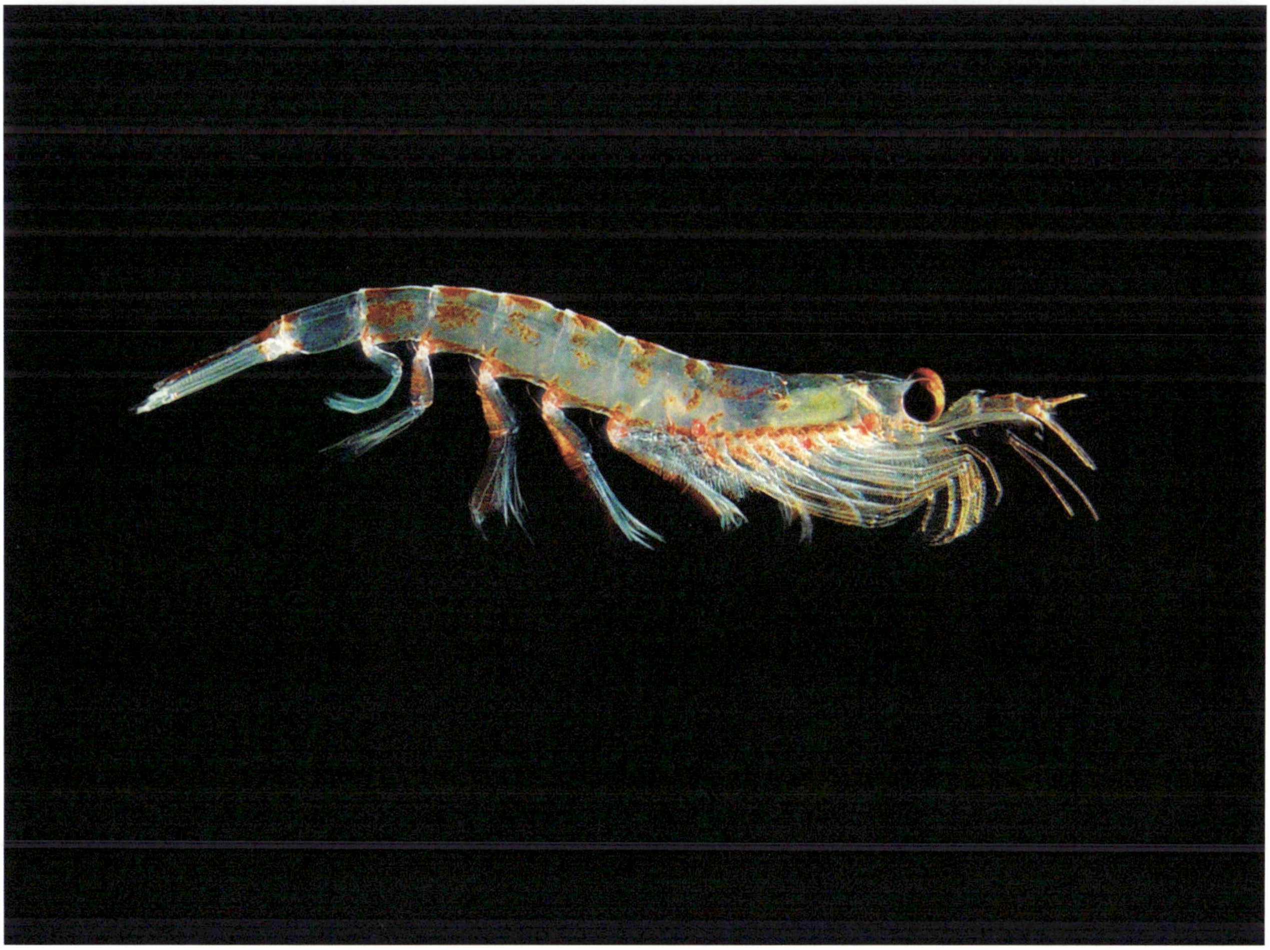

14–15

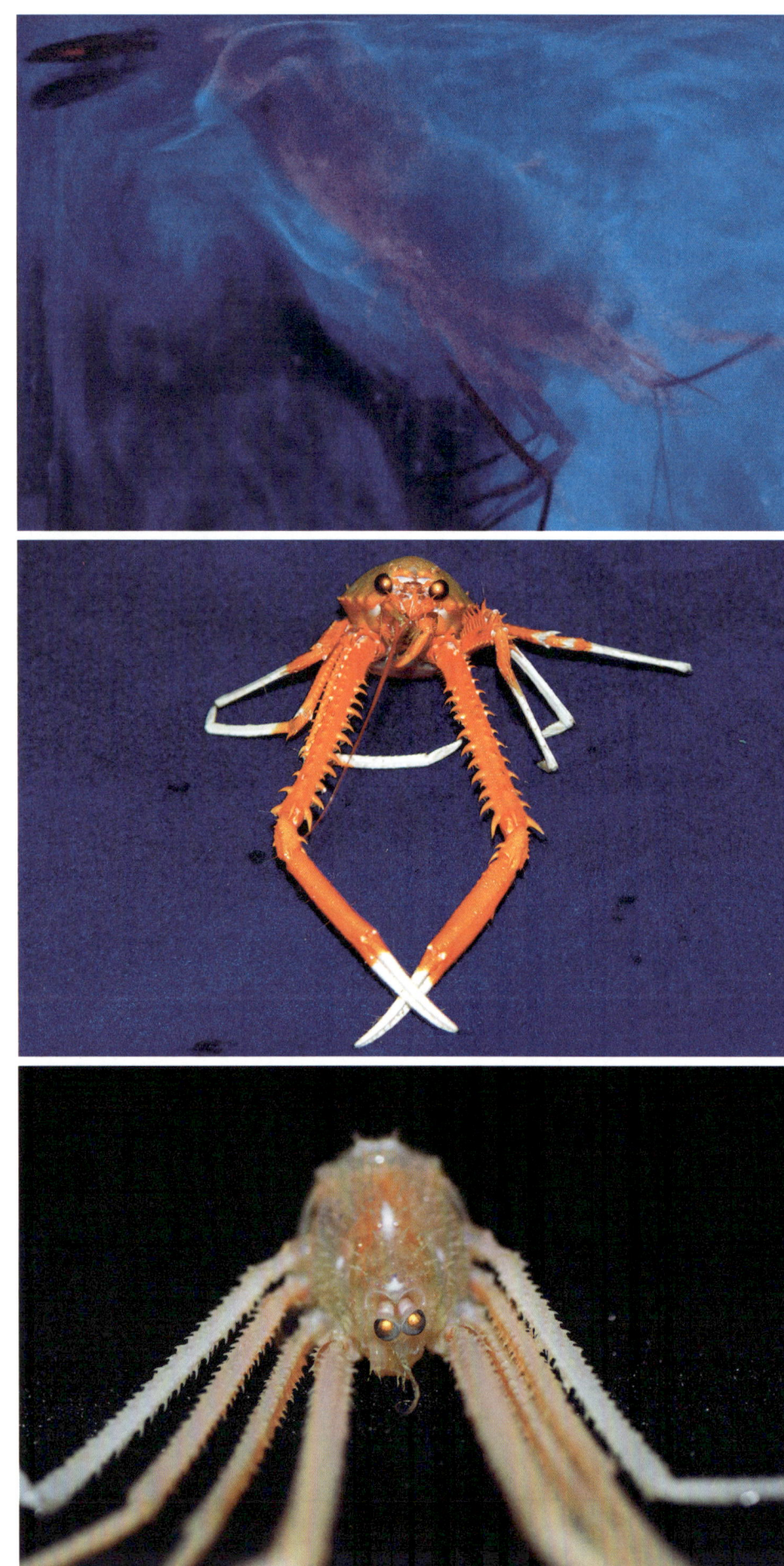

for perceiving light. Like the pelagic species, most deepsea benthic species have a single blue visual pigment, but there are two crab species (so far) that have both the blue and near-UV visual pigments [16-18]. These two species are not bioluminescent, so their dual visual pigment system must serve a different function than in the pelagic species. However, our current hypothesis is that this dual visual pigment system in these species is also related to bioluminescence, just not their own. Observations from submersibles and remotely operated vehicles by myself and other scientists indicate that these two species are often found amongst some of the "bushy" benthic animals that are attached to the bottom – gorgonians (sea pens and sea fans) [19] and anthozoans (a type of colonial anemone erroneously called the golden coral) [20].

On video footage, one frequently sees these crabs using their long arms with big claws to pick items off these structures and bring them to their mouths. Careful observations show that they are not eating the anthozoans or gorgonians, which, like many benthic sessile animals, excrete noxious chemicals to keep other organisms from settling or chewing on them. On a research cruise to the Bahamas in 2009, we were able to bring some gorgonians and anthozoans to the surface alive and were able to measure the colour of their bioluminescence. The bioluminescence from these benthic animals is somewhat more greenish than the bioluminescence found in pelagic animals. Very long-exposure photographs of *in situ* bioluminescence [21], as well as observations made from the sphere of the Johnson-Sea-Link submersible by Dr Sönke Johnsen, demonstrated that the currents were mechanically stimulating the anthozoans to bioluminesce. In addition, small planktonic animals carried in on the currents were banging into these benthic structures, which stimulated their bioluminescence as well. Our eyes can clearly distinguish between the greenish bioluminescence from the inedible anthozoans and the blue bioluminescence from the edible planktonic organisms, and it is possible that this is the purpose of the dual visual pigment system in the crabs as well. Our hypothesis is that the blue visual pigment could pick up both the blue- and green-shifted bioluminescence, while the UV visual pigment could only pick up the blue, so by processing the signals coming from these two receptor types, the crabs might be able to distinguish between the colours as well.

One other physiological adaptation found in deepsea animals to increase their sensitivity to light is a very long integration time. Much like we select a longer exposure time on a camera to take a picture in dim light (without a flash, of course), animals from dim light environments have increased the integration times of their eyes, ie, the eye collects light for a longer period of time before sending the image to the brain. The drawback of this, as we've seen with our camera images, is that the image may be blurry, but in a light-limited environment like the deepsea, some clarity has to be sacrificed to see any contrast at all. The characteristic that I use to measure the integration time is the maximum critical flicker fusion frequency (CFF_{max}). To measure this, light is flashed on the eye at increasing frequencies and irradiances, and the maximum frequency at which the response of the eye can remain in phase with the light flashes is the CFF_{max}. In our own light-adapted eyes, this occurs at about 60 Hz (60 flashes/second). Lights flashing faster than this, such as standard fluorescent light bulbs, appear as a steady glow, and the ERG no longer shows a response to every flash. Lights flashing slower than 60 HZ, such as when the fluorescent bulb starts to age,

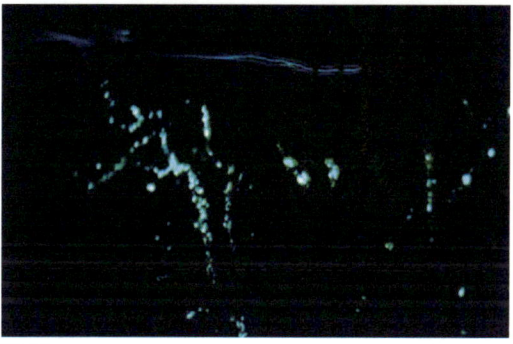

19–21

16   Natural colour image of the pandalid shrimp, *Parapandalus sp.*, emitting a bioluminescent spew. Red colouration of the shrimp is due to brief illumination by red light.
17   *Gastroptychus spinifer* and
18   *Eumunida picta*, two species of benthic crabs that possess near-UV and blue visual pigments.
19   Two *G. spinifer* sitting on a sea pen.
20   *G. spinifer* [17] and *E. picta* [18] sitting on a colonial zoantharian.
21   *In situ* photo of blue bioluminescence of planktonic animal striking a sea fan. The blue-green bioluminescence is from to sessile anthozoan, which is stimulated by water currents to bioluminescence.

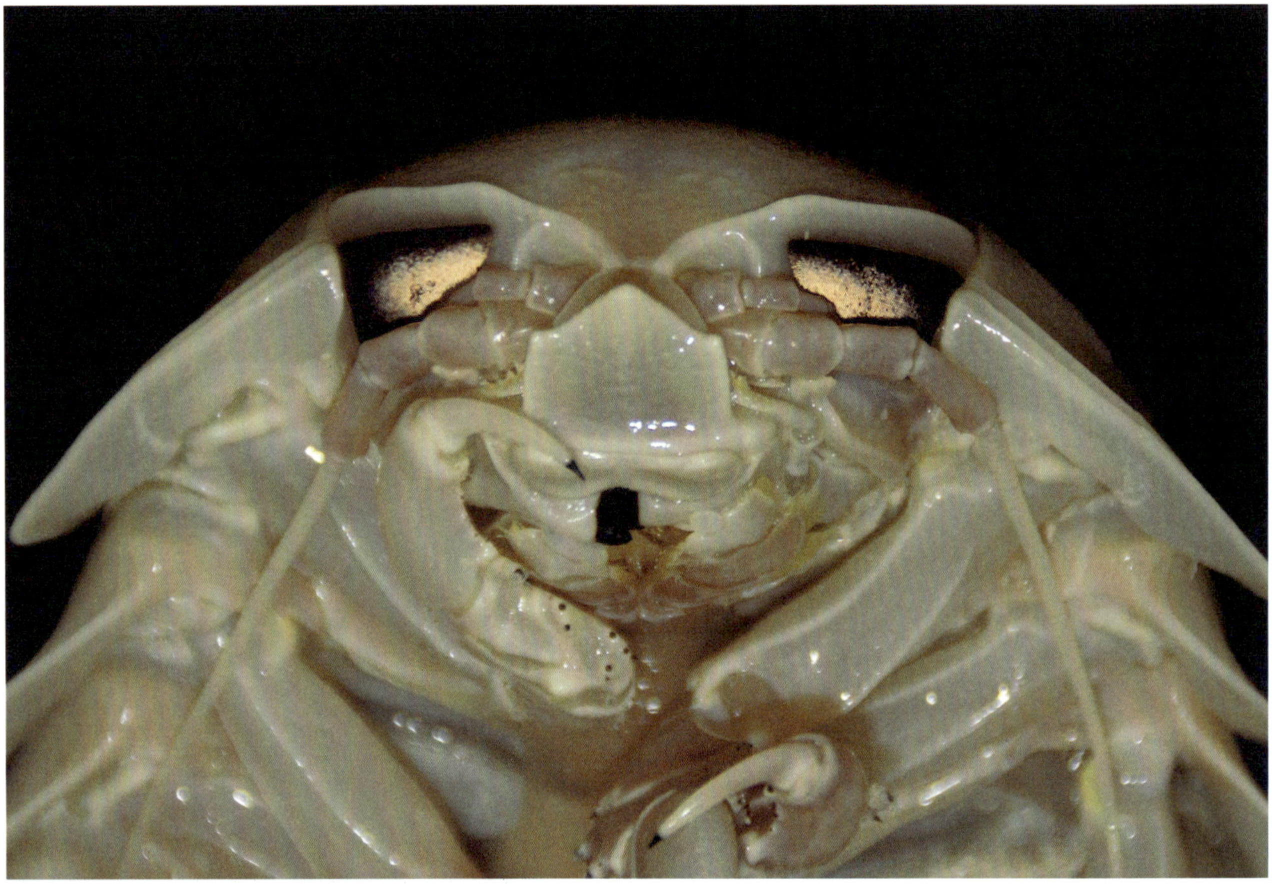

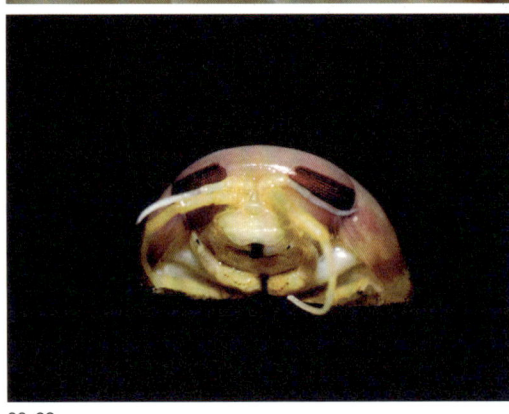

22–23

appear as annoying flickering lights, and the ERG shows a response to every flash. The reciprocal of the CFF_{max} is roughly equivalent to the integration time of the eye, so eyes with higher CFFs have a much better ability to track moving objects, but, other things being equal, have a lower overall sensitivity to light. Fast-moving animals from bright light environments have high CFFs (that of the common housefly is 200 Hz), while animals from light-limited environments, such as nocturnally active or deepsea crabs and shrimp, have flicker fusion frequencies of 15–30 Hz. However, I discovered that two species of benthic deepsea isopods, the giant *Bathynomus giganteus* [22] and its smaller relative, *Booralana tricarinata* [23], have the lowest flicker fusion frequencies ever measured in crustaceans of ~4 Hz. This means that they have amazingly long integration times, which give them excellent photosensitivity, but make their ability to track moving objects virtually non-existent. However, this may be an adaptation for their lifestyle. Deepsea isopods are thought to be scavengers, preferring to eat decaying organic matter that builds up on the seafloor. Because bacteria, many of which are bioluminescent, colonize this matter, it is possible that there are mats of dimly glowing detritus on the ocean bottom. An eye with the equivalent of a very long shutter speed, such as that possessed by these isopods, may be able to visualize this extremely dim bioluminescence, so, in this case, seeing the light would aid them in their hunt for food.

We don't know if these adaptations that we've discovered in the benthic animals are unusual or shared by a number of benthic species, because this is the first time that the visual physiology of deepsea benthic animals has been studied. However, we have an expedition in July 2015 to continue our studies of vision and bioluminescence to determine if the "colour coding food" hypothesis can be extended to other species.

22 Eyeglow from the triangular photoreceptors of *Bathynomus giganteus*, a species of isopods that can grow up to 35 cm in length.
23 The reddish strips are photoreceptors of a much smaller species of isopod called *Booralana tricarinata*.

Simon Park, molecular microbiologist

EXPLORING THE INVISIBLE

BIOLUMINESCENCE   BACTERIA   PREDATOR   SEDUCTION   INVISIBILITY
MILK SEA   EXPERIMENT

Two classic and very well known science fiction novels, which I read as a teenager, gave me an early and dramatic insight into the power and impact of the microbiological world, and reading these profoundly influenced my view of life and also my eventual practice.

The first of these science fiction novels was *The War of the Worlds* (1897) by H.G. Wells, a story that is bookended by vivid depictions of the nature and power of rarely considered microorganisms, invisible to the naked eye. At the very start of the book there is a description of the Martians regarding the Earth and its inhabitants as scientists might look down a microscope at infusoria and microbes, whilst it ends with the Earth's microbes killing the alien invaders. The second novel was Jules Verne's *Twenty Thousand Leagues Under The Sea* (1870), and there is a powerful and influential description of microorganisms in this book too. Here the book's narrator, Dr Pierre Aronnax, describes the moment when the *Nautilus* travels at night through a milk (or glowing) sea. When Conseil, his domestic servant, questions him for an explanation for the phenomenon, the following conversation, which highlights the scale, properties (although not all together correctly), and magnitude of microscopic life, begins:

> "And the whiteness which surprises you is caused only by the presence of myriads of infusoria, a sort of luminous little worm, gelatinous and without colour, of the thickness of a hair, and whose length is not more than seven-thousands of an inch. These insects adhere to one another sometimes for several leagues."
> "Several Leagues!" exclaimed Conseil.
> "Yes, my boy; and you need not try to compute the number of these infusoria. You will not be able to, for, if I am not mistaken, ships have floated on these milk seas for more than forty miles."

Verne's narrator is describing a natural phenomenon caused by bioluminescence, a process through which living organisms can produce and emit light. This biological process has arisen independently a number of times throughout evolution, and in a variety of taxonomically distinct species, in which it serves many different functions. Familiar terrestrial examples include fireflies and glowworms but bioluminescence is predominantly a marine phenomenon and the main source of light in the largest fraction of the habitable volume of the Earth, that is, its deep and sunless oceans. Its most familiar occurrence to the majority of us is when we witness the brilliant bioluminescent aurora that sometimes accompanies us when we swim in tropical seas or when we observe a striking blue luminescent bow wave or wake of a ship. In all of these instances, the light is due to accumulations of marine dinoflagellates, microscopic single-cell algae which give rise to a brief flash of bioluminescence when they are stimulated.

Verne's fictional encounter of a milk sea is based upon real events recorded in ship's logs, in which, throughout maritime history, sailors have reported travelling for hours through seas glowing with a soft white light which often extends for as far as the eye can see.

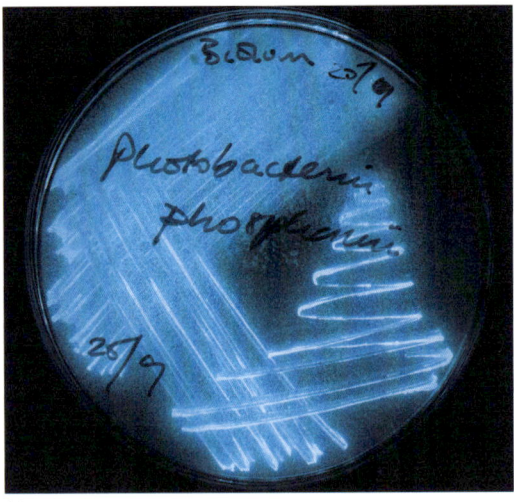

01

For many years this phenomenon defied scientific explanation, but in 2005, a study based upon data acquired in 1995 reported the first satellite observations of a milk sea. From its position above the north-western Indian Ocean, the satellite had photographed an area of glowing seawater which was some 15 400 km² in extent, and roughly the size of the county of Yorkshire. It was observed to glow for three consecutive nights, and the phenomenon was corroborated, on the first night, by a ship-based account, in which the captain reported sailing into the milk sea and then out the other side around six hours later.

The bioluminescent light that was detected from space was produced constantly throughout the period of observation and so it could not have been due to the commonly encountered bioluminescent organisms, the dinoflagellates, which are mentioned above, as these only emit light for the briefest of moments and only following stimulation. Instead, it is believed that the light was due to another type of organism with the ability to produce light, namely bioluminescent bacteria. While Dr Aronnax advised Conseil not to perform such as calculation, a conservative estimate for the numbers of bacteria necessary for the production of bioluminescence on this scale is 4×1022 (or 40 billion trillion). The scale of such a vast number is impossible for most of us to understand intuitively, but for comparison, if the entire Earth were covered in a layer of sand 10 cm thick, this is the number of sand grains that it would take to make this imaginary coating.

Bioluminescent bacteria are in fact the most widely distributed light-emitting organisms on Earth, with the majority of them living in seawater. While most species of these bacteria are fully capable of living free in seawater, a majority of them also can adopt an alternative life cycle in which they can be found living in close symbiosis

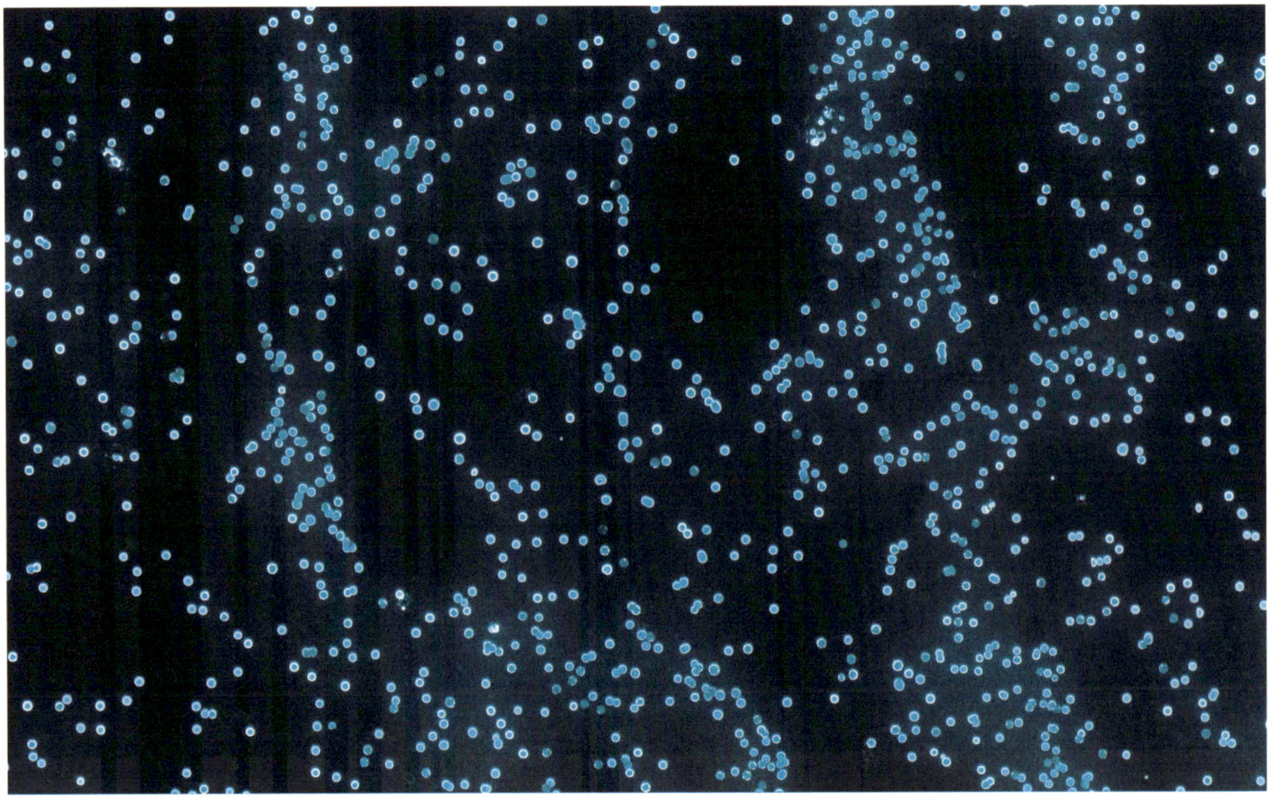

02

with larger marine organisms. In these symbioses, the bacteria are nourished, whilst at the same time the host gains the ability to utilize the adopted illumination for a number of purposes, depending upon the host species. For example, this form of bioluminescence is used for intra-species communication, as a lure to attract prey, and also as camouflage.

My first encounter with bacterial bioluminescence was within the pages of Jules Verne's novel, but my second, some eight years later, was far more intimate and ultimately more influential. As a PhD student studying molecular microbiology, I worked to isolate the genes encoding the light-producing systems from naturally occurring bioluminescent bacteria, and to introduce these into other non-light-producing and pathogenic bacteria and organisms, so that light could be used as a biomarker to track these modified organisms in complex environments. This was before the laboratory could afford expensive low-light imaging systems, so I spent many hours in the dark-room looking for the faint, tell-tale glimpses of blue bioluminescent light which meant that I had successfully transferred the genes responsible for light production into another species. Experienced intimately in this manner, and at first hand, bacterial bioluminescence is a remarkably powerful and refined light. It is cold but also uniquely beguiling, and as perhaps fits one of its major biological roles as a lure, it attracts and engages the curious. For these reasons, and for many years now, I have used bacterial biolumines-cence in my practice beyond scientific research. My long-standing investigation into the aesthetics of bacterial bioluminescence, and how I have used this organic form of light as a unique medium to disclose some of Nature's most vital yet often unseen events, has involved bioluminescent photo booths, old cups, saucers, glasses, spoons, bowls and vases purchased from charity shops, and has led to collaborations with artists and scientists from different fields.

Beyond its ability to directly engage people with its captivating cold glow, bacterial bioluminescence also has many applications in medical and biological research, and because light is so easy to detect, it is very widely used as a biomarker. An early historical example of such an application was when bioluminescence was observed in 1825 in two discarded corpses at The Anatomy School in London. When the luminous microorganisms were scraped from the corpses it was found that they could make others glow too, providing one of the very first practical demonstrations of the role of microorganisms. More recently, bioluminescence imaging has enabled disease-causing bacteria, genetically modified to produce this light, to be tracked during infection in a noninvasive manner in living animal models. In a sense, then this unique light of biological origin allows us to explore and illuminate otherwise invisible processes and events, and this concept has very much become part of my own artistic practice.

Humans have known about bioluminescence for thousands of years: it appears in the folklores of many countries and long before any proper scientific explanation could have been offered to explain it. It is often encountered as an ephemeral and unexpected phenom-enon, and it has an ineffable power, acting to attract and engage the curious, as befits one of its biological roles as a lure. In my work I explore "the invisible", but it has engaged and inspired curiosity in far greater individuals than myself, and I am certain that it will continue to do so in the future. And now we have the expertise to submerge all the processes of life on Earth in bioluminescence.

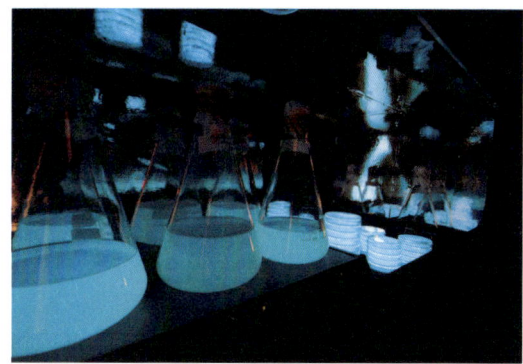

03

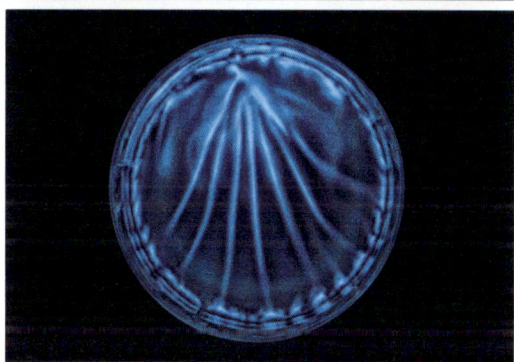

04–05

01–02   The marine bioluminescent bacteria *Photobacterium phosphoreum* grown on biolum-inescent agar and imaged in the dark. The bacterium streaked on to agar, and growing in a conventional Petri dish [01], and inoculated on to a large, square plate [02] to obtain single colonies. In the latter image, each pinpoint of light is a colony of the biolum-inescent bacterium, each one containing around three billion light-producing bacterial cells.
03   The inside of the Bioluminescent Photo-booth. This was a self-contained photo booth, installed at the Royal Institution in 2008, in which the only sources of light for portraiture were large flasks, and hundreds of Petri dishes, containing cultures of *P. phosphoreum*.
04–05   Storm in a teacup and wineglass. Liquid cultures of *P. phosphoreum* in a teacup [04] and a wineglass [05]. The photographs show the complex and ever-changing patterns in the bioluminescence that were observed. Whilst there is no proper scientific explanation yet, the patterns most likely reflect changes in the concentration of oxygen.
06   [next pages] The Bacterial Light Lab. Redundant laboratory glassware was filled with bioluminescent bacteria. This striking installation was exhibited at the On Light event at the Wellcome Collection in 2015, where the lure-like quality of the bioluminescent light attracted and captivated many visitors.

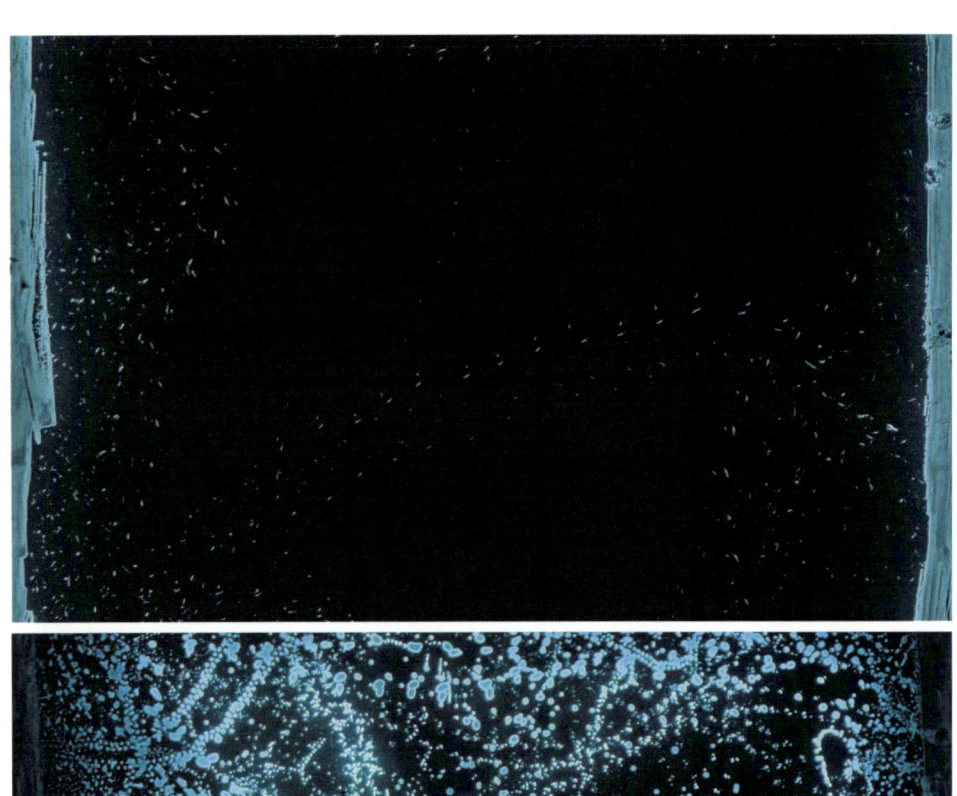

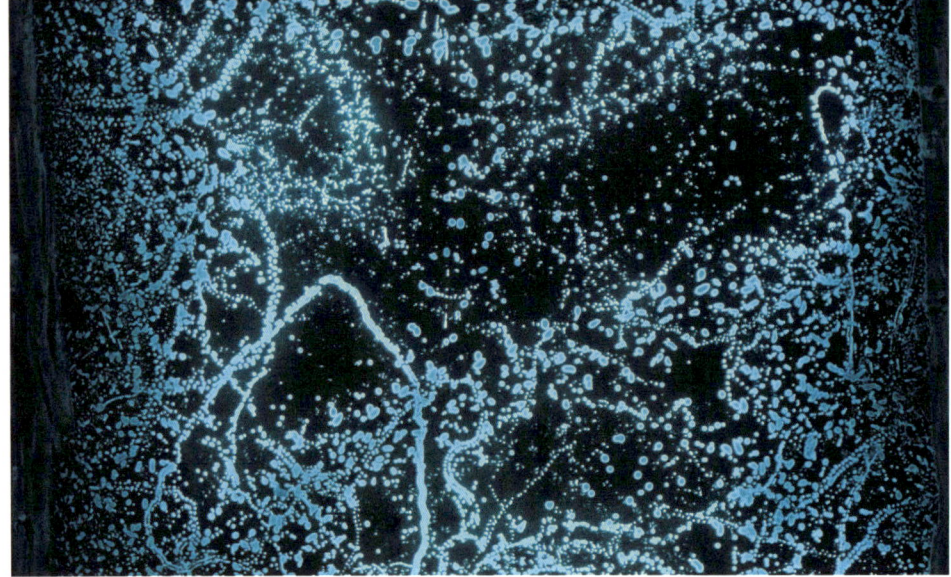

07–08

Soil teems with near-microscopic life in the form of small animals called the cryptozoa. These are minute invertebrate animals, just about large enough to be visible to the naked eye, which live between litter and the soil, and which are vital for the health of soil, and for the breakdown and recycling of the litter material. Drawing on the theme of exploring the invisible, and the medical and biotechnological uses of bioluminescent bacteria, I have developed a process which for the first time reveals the usually overlooked but myriad activity of these minute creatures. Bioluminescent bacteria are inoculated on to the surface of a device containing agar so that they cover the circumference of the medium. The device is then implanted on to the surface of the soil and observed over a period of up to four days. When the cryptozoa enter the device, they inevitably walk over the bioluminescent bacteria and inadvertently collect them on their feet (or other bodily parts,

depending on their means of locomotion). Then, as they continue their journey and walk over the uninoculated agar surface, they leave behind a trail of the bacteria in their footsteps. Because of the microscopic nature of the footprints, and the small initial numbers of bacteria, these tracks are at first invisible. However, after a day or so the bacteria grow into visible points of light, which reveal the otherwise invisible tracks left behind by these creatures and provide an insight into the vastness and complexity of the activity of the cryptozoa [07–08]. In a sense, the growth of the bacteria, and their production of light, acts as an amplification process to reveal what is normally invisible. This process also gives an excellent and direct visual assessment of the biodiversity present and the health of a soil, as the track-patterns formed from intensively farmed soils are far less complex than, for example, those from old woodlands and ancient non-agricultural environments.

I have used bioluminescent bacteria to provide a direct visualization of the process of photosynthesis for the first time. This unique process relies upon the mechanism that the bacteria use to make bioluminescent light. These bacteria possess a light-generating enzyme called bacterial luciferase, and its substrates are a reduced flavin mononucleotide ($FMNH_2$), molecular oxygen and a long-chain fatty aldehyde. Luciferase catalyzes the synthesis of a fatty acid from these reactants, and the excess energy, which is liberated from the oxidation of $FMNH_2$ and the aldehyde, and the concomitant reduction of molecular oxygen, is released as a photon of blue / green light. During the normal growth of the bioluminescent bacteria in the laboratory, oxygen is usually the limiting factor, and if the bacteria are starved of this element whilst they are still alive and capable of growth, they will no longer produce light and the culture will become dark. In this state, a culture of *Photobacterium phosphoreum* is uniquely receptive to oxygen. Photosynthesis is the process that plants use to convert light energy, carbon dioxide and water into chemical energy that can be released to fuel the organisms' activities. In addition, oxygen is also produced as a "waste" product of this reaction, and so the generation of this element here becomes a detectable signature for photosynthesis. Consequently, in order to observe photosynthesis, a culture of *P. phosphoreum* was carefully starved of oxygen so that it became a unique medium able to respond to and detect this element. A plant was then carefully submerged into the liquid and the culture kept in the dark for 60 minutes so that any residual oxygen was removed and photosynthesis was stopped. Next, to initiate photosynthesis, the plant was illuminated for a brief period, and then the light turned off so that any oxygen liberated by the photosynthesis reaction would be revealed by its activation of bioluminescence in the culture of *P. phosphoreum*. What emerges from this process is an image of a plant that is surrounded by what appear to be blue flames [09], and which in reality is a direct and unique representation of the oxygen that it has liberated during its brief period of photosynthesis.

09

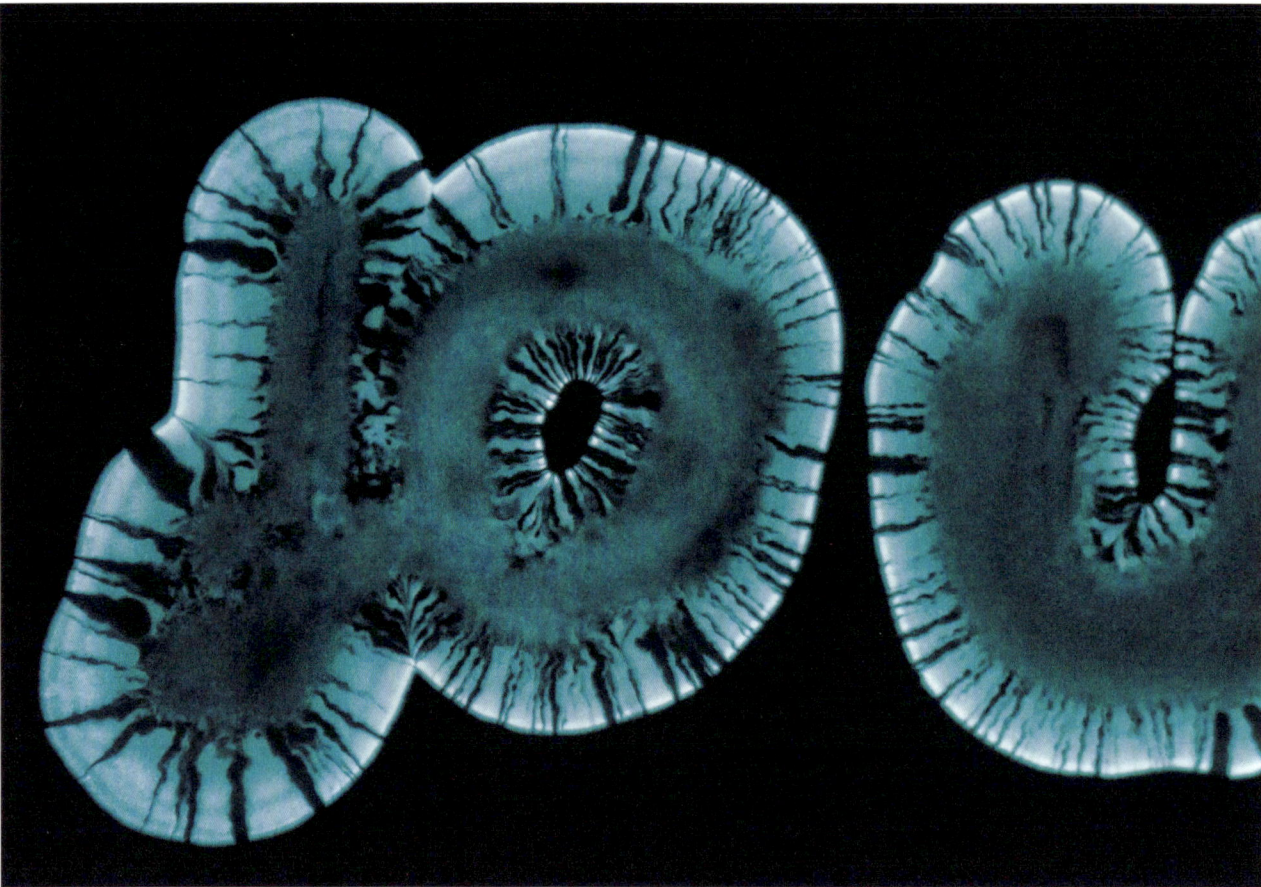

10

The final exploration, in which bacterial bio-luminescence is used to reveal usually unseen yet important natural mechanisms focuses on perhaps the most important one of all, and the one that is the basis of all life here on Earth, that is the process of evolution itself. Unlike humans, who are diploid and contain paired chromo-somes, one from each parent, bacteria have only a single-copy DNA genome. Any changes in the DNA sequence of their genome results directly in measurable phenotypes, as there is no blending or filtering of the informational content of the DNA, as is the case in higher organisms. Because of this, mutations or glitches in the DNA code can become readily visible in the characteristics of the bacterium in which these have occurred. The bioluminescent bacteria that I use in my work naturally produce bioluminescence, but from the perspective of the bacterial cell, this process is metabolically expensive, because it utilizes vital resources that could have alternative use elsewhere within the cell. Any bacterial cell that loses the ability to generate light will be able to grow more efficiently and thus faster than cells that do produce it. In the absence of a Darwinian selective pressure then, there is a tendency for cells of bioluminescent bacteria to lose the ability to produce light through spontaneous genetic mutation and variation, the same processes that power biological evolution. In the process that I developed here, bioluminescent bacterial cells were carefully prepared and used as a printing ink. When printed on to an agar surface, not only do the bacteria grow to form a unique living and glowing text, they also reveal the process of evolution at work. The dark striations that can be seen within the blue bioluminescent bacteria [10] are generated by spontaneous genetic mutation within the genomes of the bacterial cells. These new dark variants no longer possess the ability to produce light, which on the laboratory agar is a burdensome and unnecessary metabolic process. These dark cells are thus starting to segregate from the light-producing parental bacterial strain, and therefore are beginning an evolutionary journey into a new, lightless species.

Lex Kaper, astronomer

DETECTING DISTANT STARS AND GALAXIES

COSMOS STARS TELESCOPE INNOVATION SPECTROGRAPHS
DARK MATTER INVISIBILITY

Light carries information on the origin of the universe, the evolution of galaxies and the physical nature of the stars and the planets. Stars are the only objects in the universe that produce (visible) light. Thanks to the stars the dark universe is illuminated, all chemical elements besides hydrogen and helium are produced, surrounding planets are warmed up and life has been able to emerge, at least here on Earth.

The light produced by the sun is the most important ingredient for life on Earth. But light is also the messenger carrying information on the content of the universe. By detecting their light we can study distant stars and galaxies that are way beyond our physical reach.

The sun is our nearest star. Although light travels with an enormous speed, 300 000 km/s, it still takes eight minutes to travel from the sun to the Earth. Given the enormous distances between the stars (the nearest star, Proxima Centauri, is at a distance of four light-years) and towards galaxies (millions to billions of light-years), one needs big telescopes and sensitive instruments to detect the light emitted by the stars.

The telescope is a Dutch invention: in 1608 the Dutch optician Hans Lipperhey visited Prince Maurits, the Viceregent of Zealand and Holland, in The Hague, where he demonstrated the telescope [01].

01

This event is reported in the newsletter *Ambassades du Roy de Siam envoyé a l'Excellence du Prince Maurice, arrivé à La Haye le 10 Septembre 1608*. The newsletter reports on the first visit of a Siamese diplomatic mission to Europe, the peace negotiations with Spain *and* the aforementioned demonstration of the newly invented telescope. This was almost one year before Galileo Galilei in Italy was able to discover the moons of Jupiter with an improved version of the telescope. In 1568 the Dutch Republic was at war with the King of Spain, Philip II. The Eighty Years' War would last until 1648, but was interrupted by the Twelve Years' Truce, which started in 1609. In 1608, the commander-in-chief of the Spanish forces, Ambrogio de Spinola, First Marquis of the Balbases, was in The Hague to represent Spain in the peace negotiations. The Dutch East India Company (VOC),

01   A small telescope dating from c. 1625, excavated in Delft in May 2014. It is similar to the one produced by Lipperhey in 1608, and barely worked as well as the now commonly used binoculars. However, it enabled one to see distant objects, invisible to the naked eye, such as the millions of stars in the Milky Way.

02

the world's first multinational, was setting up a trade mission in the small Malay state Patani (today a southern province of Thailand), considered by the Dutch as the entrance to Siam and to China. The Portuguese had started a rumour that the Dutch were buccaneers without a country of their own, and so the Dutch welcomed the initiative of the Siamese King Ekathotsarot to send a mission to Holland. The newsletter reports that, "A few days before the departure of the Marquis Spinola from The Hague, an optician presented His Excellence with a philosophical toy with which it is possible to see the windows of the church in Leiden. The glasses are very useful to inspect objects at a distance of a mile and further, as if they are very nearby. Even the stars that are usually invisible due to their small size and our poor sight, can be seen with this instrument."

Since 1608 the diameter of the primary mirror / lens of the telescope has grown exponentially with time. Sir William Herschel, discoverer of Uranus in 1780, built fantastic telescopes with apertures exceeding one metre. His famous 40-foot telescope was built in Slough, England, in 1789 and used a 120 cm diameter primary mirror with a 12-metre-long focal length. It was the largest telescope in the world for 50 years. In the early 20th century the first telescope with a 2.5 m mirror was constructed, the *Hooker* telescope of Mount Wilson Observatory used by, among others, Edwin Hubble. With the construction of the 5 m *Hale* telescope in 1948, the maximum size of the conventional reflector had been reached. Just before the millennium change, alternative designs (segmented mirror, supported meniscus mirror) resulted in the first 8–10 m class telescopes (eg, the *Keck* telescopes and the European Southern Observatory's *Very Large Telescope*, [02]). Although we are still exploring their scientific potential, the next generation of 30 m telescopes (*Giant Magellan Telescope* (GMT), the *Thirty Metre Telescope* (TMT) and the 39 m

European Southern Observatory's *Extremely Large Telescope* (E-ELT) is planned to see first light[1] in the next decade. Since January 2016 I have been in charge of the development and construction of MOSAIC, a multi-object spectrograph, which will be installed in the E-ELT on the peak of Cerro Armazones in Chile by 2024.

Obviously, it is not only the telescope's primary mirror diameter that counts; the available suit of instruments also determines for a

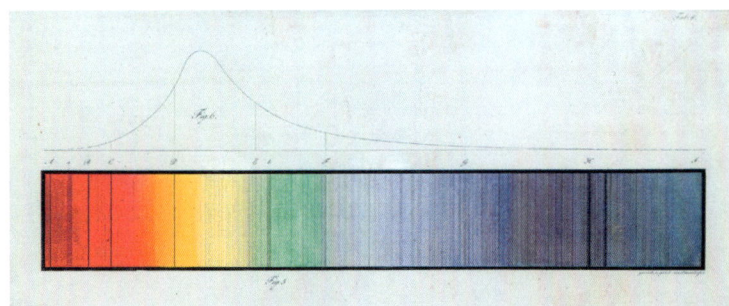

04

large part the scientific power of the light-observing facility. Some instruments are used to make images of stars and galaxies, using various broad- and narrow-band filters to measure the objects in different colours. Other instruments, called spectrographs, make a spectrum. Using a prism or grating, the light is dispersed into its colours. When using a slit, absent colours become manifest as dark lines in the spectrum [03]. These lines are produced by different chemical elements in the atmospheres of the sun and other stars, ie, light of specific colours, or wavelengths, is absorbed by atoms and ions. As these "missing" colours are specific to the atom, one can determine their presence from the pattern of lines in the spectrum (like a barcode). In Fraunhofer's first spectrum of the sun, the element helium was discovered (named after *helios*), which is hard to detect (though present) on Earth.

Presently, the most powerful spectrograph in the world is X-shooter on the *Very Large Telescope* (VLT). My role in this project was to set up an international consortium of astronomers from Denmark, France, Germany, Italy and the Netherlands, who were selected by the European Southern Observatory in 2003 to design and build this state-of-the-art spectrograph. The X-shooter consortium includes both scientists and technicians, so that the scientific requirements result in the right technical design of the instrument. X-shooter was mounted on Unit Telescope 2 of the VLT in 2009. A unique feature of the spectrograph is that it covers the full optical and near-infrared wavelength range (from the atmospheric cut-off at 300 nm up to the K band ending at 2500 nm). The large wavelength coverage provides the opportunity to identify the physical nature of any unknown object "X".[2]

Astronomical instrumentation is very specialized and unique in nature, so one cannot just order such a spectrograph from industry. The process of designing and optimizing the optical and mechanical design takes several years [04]. As the X-shooter is mounted in the Cassegrain focus of the telescope (ie, below the hole in the primary mirror of the telescope), it moves with the telescope when guiding on to a target [05]. This makes weight and flexure critical parameters in the mechanical design of the spectrograph. Therefore, the spectrograph housing is extremely stiff, but also very lightweight, and adjustable mirrors are included to correct for motions in the spectrograph due to flexure.

03

02   The Very Large Telescope at ESO Paranal in the Atacama desert in Chile. It is located on a mountaintop of 2650 m in one of the driest regions in the world. Bad conditions for life, but almost ideal conditions to observe the sky. The four 8.2 m telescopes carry various instruments (the "eyes" of the telescope) to detect the light emitted by the stars and galaxies.
03   Mechanical design of the VLT/X-shooter spectrograph. Light from the telescope enters the instrument at the top and is split into three separate spectrograph arms using dichroics. These optical elements are fully transparent for light with wavelengths longer than a given value, and totally reflective for light of shorter wavelengths. The wavelength coverage of X-shooter is unprecedentedly large, including the ultraviolet down to the atmospheric cut-off at 300 nm due to ozone up to the near-infrared ending at 2500 nm.
04   Spectrum of the sun obtained by Fraunhofer in 1814. The dark lines ("absorption lines") are caused by atoms, ions and molecules in the solar atmosphere. Some of the darkest lines are actually formed in the Earth's atmosphere. At the top the spectral energy distribution of the sun is shown, peaking at 500 nm (yellow). That is the reason why the human eye is most sensitive to yellow light.

The first X-shooter spectra were obtained in 2009 [06] during the commissioning period of the instrument. The X-shooter consortium received guaranteed observing time to compensate for the costs and effort of building the instrument. My research group at the University of Amsterdam used that time to study the formation of massive stars deeply embedded in their natal clouds, to obtain spectra of massive stars in galaxies in and even beyond the Local Group (the galaxy

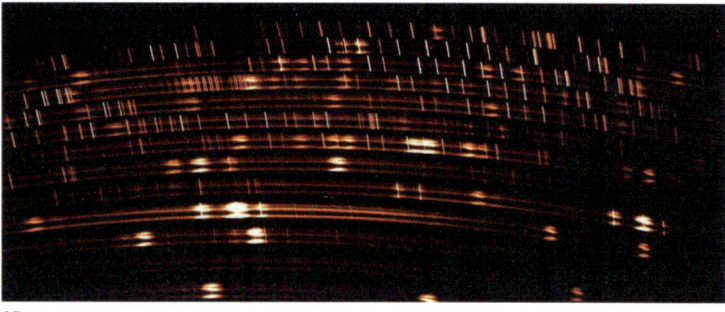

05

group that includes the Milky Way), and to detect the optical and near-infrared afterglow of gamma-ray bursts: exploding massive stars leaving a black hole at distances of several billion light-years.

Since the installation of X-shooter on the VLT it has become one of the most demanded instruments in the world. Over the first five years of its operational life, more than 200 scientific X-shooter publications have appeared in professional literature, covering topics ranging from the formation process of stars, very old, metal-poor stars in the galactic halo, the properties of massive stars in Local Group galaxies, the intergalactic medium as probed by the sightlines to distant quasars, to star-forming galaxies at the edge of the visible universe.

All these projects have in common that the light emitted by these distant objects provides information on the physical processes acting in outer space, increasing our knowledge of the laws of physics that became apparent when doing experiments here on Earth, but that can be further tested and explored under the extreme conditions present elsewhere in the universe. Unexpected discoveries, such as dark matter and dark energy, tell us that the universe is much more diverse, complicated and interesting, than we thought on the basis of terrestrial experiments.

Although astronomy is a fundamental science aimed at understanding our cosmic origin and position in this expanding universe, the development of technologies to build sensitive astronomical instrumentation has led to many important contributions to society. For example, the glass used to produce the big mirrors of telescopes, known as Zerodur, has the property that it does not expand or shrink when changing its temperature. Such glass is now commonly applied in ceramic hobs. Prosaic perhaps, but useful nevertheless. Astronomical objects also emit light invisible to the naked eye, such as X-rays or radio radiation. Satellites launched by rockets in the 1960s detected the first X-ray sources in the sky. The development of these X-ray satellites has resulted in the application of X-ray detectors at airports and in medical instruments. The techniques used in radio telescopes are now applied worldwide in WiFi networks.

Astronomy is arguably the oldest science in human history. Light has always been central to our scientific progress: from the observation (and accurate prediction) of solar eclipses by Babylonian astronomers, to the discovery of dark matter and dark energy not producing light and demonstrating that we "only" understand the

05   Image showing the echelle spectrum of supernova 1987A detected by the visual arm of VLT/X-shooter. The X-shooter slit was oriented along the ring-like structure of the supernova remnant resulting in the two bright stripes per echelle order. The rings produce strong nebular emission as can be seen from the bright spots appearing in the image. The tilted straight lines are produced by the Earth's atmosphere.
06   The X-shooter spectrograph mounted at the Cassegrain focus of one of the four unit telescopes of the ESO *Very Large Telescope*.

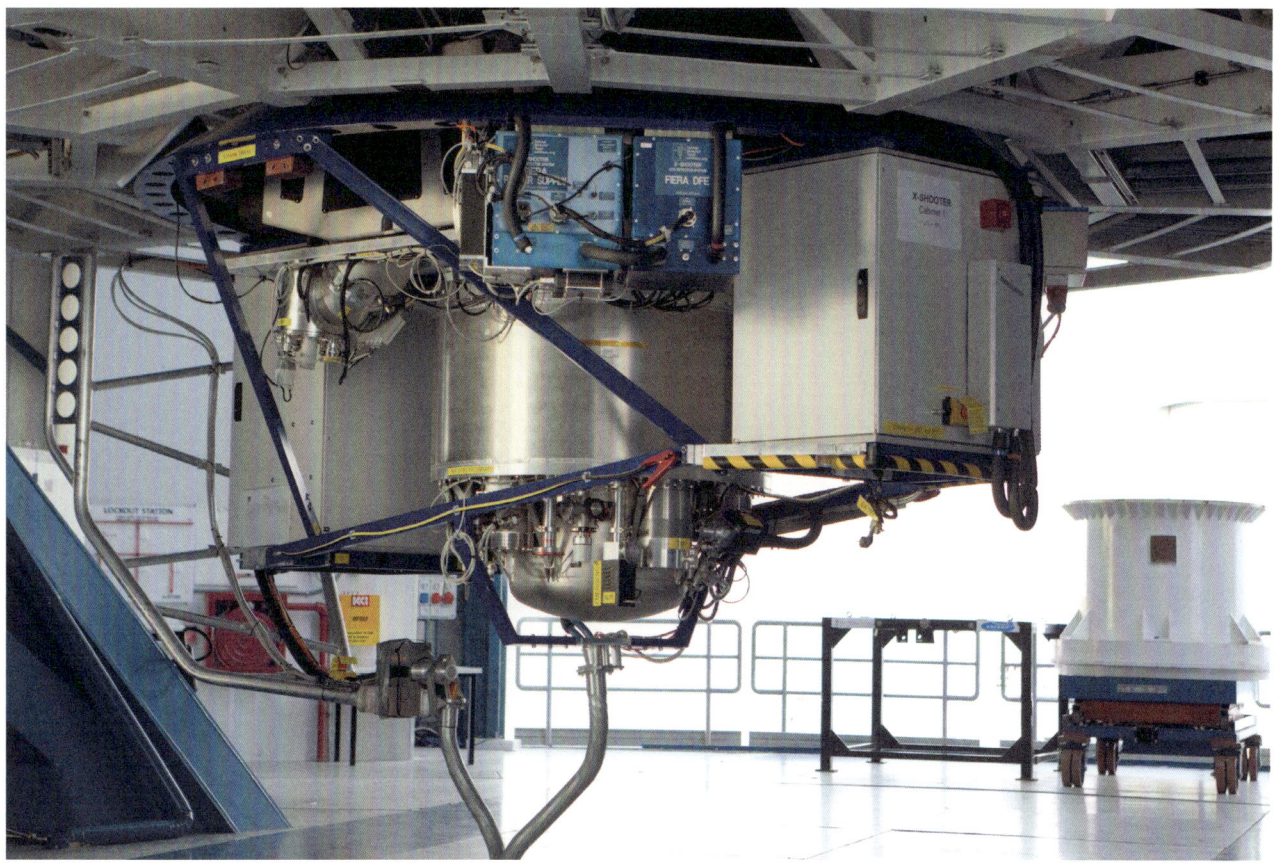

05

nature of just 4% of the matter and energy in our universe. It is a privilege to be part of the current "Golden Age" of scientific progress with many important discoveries ahead in the coming decades, addressing questions like: what is our universe made of, and how did it evolve? Are there planets like the Earth that managed to produce life? And, are we alone?

1   Object X stands for the (non-existing) tenth planet in our solar system.
2   First light is the term used to mark the first use of the telescope.

David Peggie, analytical chemist

SCARRING WITH LIGHT
PART 1

DECAY PIGMENT PRESERVATION EXPERIMENT

As a conservation scientist working in a museum, my scientific knowledge is usually applied to understanding "old" works of art; objects created many years ago and which have gradually changed over time. I usually come to an object when the damage has already been done, but for "Scarring with Light", Natalia Zagorska-Thomas asked me to help her create an object that would be deliberately destroyed by light.

Brazilwood has been used for centuries to make red dyes and pigments. Brazilwood-derived colourants fade quickly when exposed to light. The propensity of these pigments to fade makes them very difficult to identify in specific paint passages, especially when these passages contain a complex mixture of pigments. However, recent work has enabled identification of brazilwood-derived pigments in paintings by artists such as Rembrandt. I have studied these dyes and pigments in laboratory experiments over the past decade, trying to understand their precise degradation mechanisms and to find ways to identify their faded remains. To detect and investigate these faded pigments I use an array of spectroscopic techniques, many of which harness the properties of specific wavelengths of light as they interact with matter. So, while light may have a role in causing damage, light also becomes the means by which the damage can be investigated.

The use in historical artworks of pigments or dyes that fade easily when exposed to light can make interpreting an artwork problematic, and in some instances may even affect the longevity of the work itself. In contrast, for this project Natalia and I actively selected this extremely fugitive dye; an undesirable quality in most circumstances, but here a positive attribute. Areas of the brazilwood-dyed silk would be exposed to light, creating an image from the destruction.

Performing the dyeing in Natalia's studio was surprisingly exhilarating. This arose from the scale of the procedure; the dyeing was done on large pieces of woven silk rather than on a small sample of thread, as is more usual when conducting laboratory experiments. But it also arose from something else, something more fundamental. Somehow the

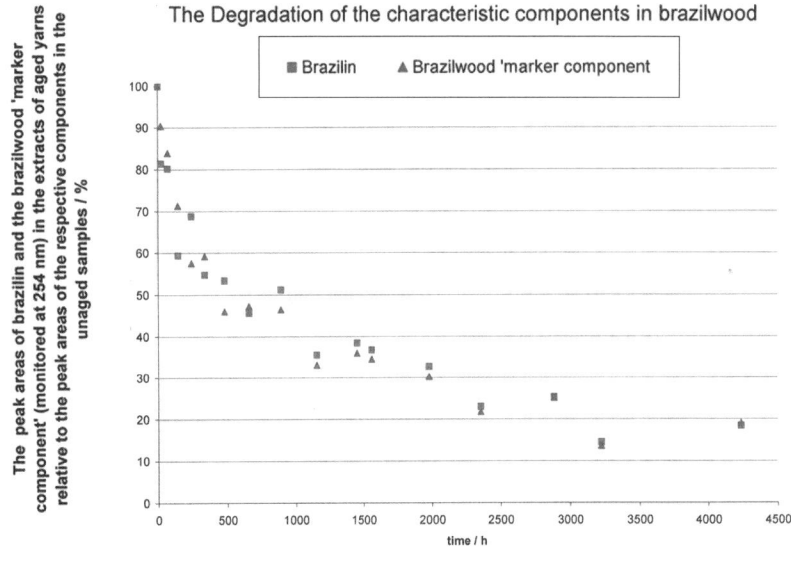

01   This graph shows the degradation of two distinct chemical components found on yarns dyed with brazilwood. The photo-degradation rates of the two components are identical and c. 20% of both components remained after approximately 25 weeks (4200 hours) exposure in an accelerated ageing box consisting of 12 × 20 W fluorescent tubes.

rules had changed and I was no longer testing the fading properties of a particular dye component or investigating how it might be detected, but helping to create an artwork. Seeing the bright-red fabric spilling out of the kitchen pan and knowing that it was to be used in the production of a real object, I began to get a sense of how the ancient dyers must have felt, transforming undyed cloth into colourful starting materials to be used by artists in everything from clothes to decorative objects.

During the anxious wait for the destructive action of light to create the work, I unexpectedly began to feel "responsible" for the science, worrying that maybe it wouldn't fade; maybe "real life" wasn't like my laboratory experiments and the textbook descriptions. These predicted that the brazilwood-dyed silk would fade appreciably in around five weeks. But for the first stressful two weeks nothing appeared to have happened at all.

Our collaboration allowed me to investigate the science of destruction in a playful way, to help produce an artwork based on the intrinsic fragility of brazilwood dye.

Natalia Zagorska-Thomas, artist, textile conservator

SCARRING WITH LIGHT
PART 2

PIGMENT   EXPOSURE   DESTRUCTION   EXPERIMENT   TIME

These are the beginning stages of a long-term, collaborative art project between David Peggie, a conservation scientist, and myself, an artist and a textile conservator. Having often discussed the complex relationship between art and science within the heritage preservation industry we wanted to devise a project where we could explore this interdependency in a more open, creative way. We came upon an idea, driven largely by my art practice, for a series of artworks made by controlling the amount of light exposure upon a natural dye to produce an ever-changing image in a way similar to a very long-exposure photograph. The bleaching effects of light upon an artefact over time are both the subject and the means by which images are being slowly produced on a silk dress dyed using a traditional recipe involving the core of a redwood tree.

Conversely, my job as a conservator involves both counteracting the effects of light damage on objects and preventing it from happening in the future. My professional code of ethics is based on custodianship, the notion that cultural material belongs to all and is held in trust on behalf of present and future generations. Ultimately of course, the project is doomed to failure: as a conservator I can only slow down the effects of time and the environment on organic material – never arrest it completely. Both as a conservator and as an artist, it seems to me that to eradicate decay completely would be to deny history, the passage of time and the inevitability of change, without which nothing new can be created. It is not an uncomplicatedly happy thought, and yet, somehow there can be relief, joy even, in realizing that we can't have ultimate control over our material culture any more than over the span of our own lives.

While working on many faded and degraded textiles as a conservator, the artist in me has become aware of the immediate, arresting beauty and the metaphorical potential of scarring created by light damage. When examining the colour variations within the folds of a Victorian curtain or the lining of a Renaissance tapestry, one becomes aware of new patterns in these fading objects created by the slow exposure to light. Exposure can take anything from a few weeks to several hundred years. In addition to bearing witness to the specifics of passing time, the damage often exposes previously invisible physical structures that tell a totally new and unexpected story, one, which, as a conservator, I must be very careful to eke out and present according to strict ethical rules and factual evidence.

As an artist, however, I am under no such constraints, I can tell whatever story I choose. It may seem that my work as an artist and as a conservator are in direct opposition, but to me they appear rather as two strands of dealing with the same fact as glaringly obvious as it is ungraspable: nothing stays the same.

BRAZIL

Our choice of Brazilwood-dye for this project came about as a result of its known high susceptibility to light damage. The dye derived from Brazilwood shares its name with the South American country. It would be logical to assume that the tree and the dye are named after the land whence it came, but the truth is exactly the reverse. The word Brazil is derived from the medieval Latin *lignum brasilium* (or *brisilium*), ("red like an ember", from the medieval Latin *brasa*, "ember" or old French *brese* and Germanic *brasa*), referring to the reddish/orange colour obtained from European redwood trees harvested and used for colouring purposes since the middle ages. The country gets the name by which it is now known from 16th-century Portuguese arrivals who turned the abundant forests they found there into one of the first massive-scale colonial exports. The story of Brazilwood is a story of discovery, commerce and trade. It is also the story of conquest, enslavement and environmental destruction. The trade continues still, though it may soon become illegal due to the near extinction of the Brazilian redwood forests. The very quality that is considered a weakness in conservation became a strength in this art project, where we rely on the bleaching of colour to produce an image.

THE EXPERIMENT

Aim: To find out what colours can be obtained with Brazilwood using different additives during dying (acid vs alkali).

To produce the dye extract: Boil 100 g of Brazilwood powder in 1600 ml of water. This should leave you with 1.5 l dye.

To prepare the fabric: Pre-boil 500 g of habutai silk to remove any industrial finish, which might prevent uptake of colour.

Alkali batch:
1 Mordant* 250 g of silk with 5 g of alum.
2 Add 1.5 g of cream of tartar to mordant for each 50 g of fabric.
3 Place 250 g in 250 ml of dye extract with approx. 4 g of potassium carbonate in 2 l of water.
4 Bring to the boil. Continue for 30 min.

Result: Silk became a deep-pink colour.

Acid batch:
– Mordant 250 g of silk with 5 g of alum
– Add 1.5 g cream of tartar to mordant for each 50 g of fabric
– Place 250 g in 250 ml of dye extract with 2 tbsp of lemon juice followed halfway through the cycle with a pinch of powdered chalk.
– Bring to the boil. Continue for 30 min.

Result: The fabric became a very bright yellow colour as a result of the acid in the lemon juice. A pinch of powdered chalk (alkali) brought the colour back slightly towards the blue end of the spectrum, producing a salmon-pink colour.

Mordant: from French "to bite". A chemical substance used to set (attach) dye onto the fabric by boiling fabric with the mordant substance, in this instance alum, before starting the dye cycle.

EXPOSURE TO LIGHT

Aim: To ascertain how quickly habutai silk fabric dyed with a traditional brazilwood recipe succumbs to the effects of light.

Small rectangular samples of dyed silk from the two batches described above were exposed to synthetic light from a 70 W 580 lumen halogen bulb for five weeks, followed by three weeks of exposure to natural light through a plate-glass window without a UV filter.

Result: Fading was the same for both batches.

Artwork: Using the recipe which produced a more salmon- coloured sample, I cut the silk in the shape of a dress, leaving a section exposed to represent the scar-like Rio Grande do Norte tributary in Brazil where brazilwood was originally harvested. As both colour samples faded at the same rate during the first phase of the experiment the choice of shade was

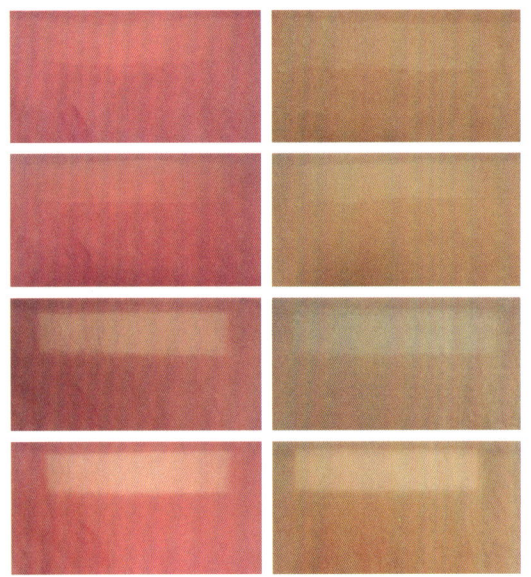

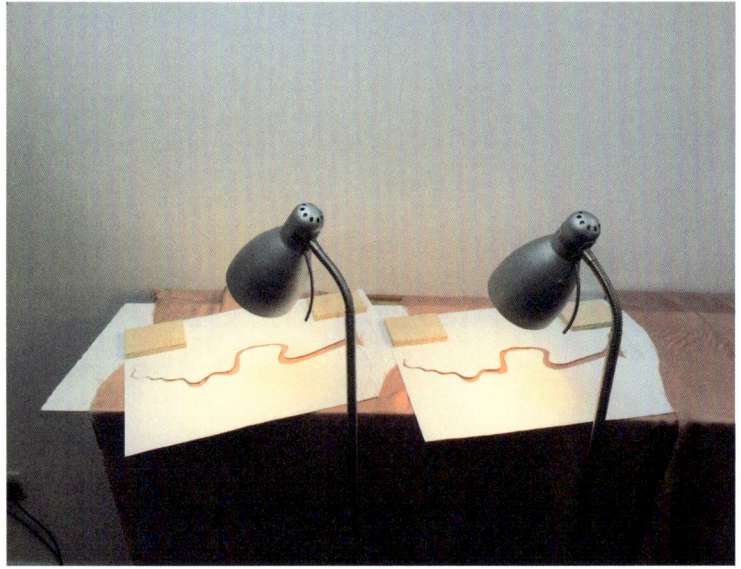

driven purely by the concept of the piece. This colour is more redolent of skin, which fitted with the idea of scarring upon the body of the dress. The scarring effect was created by exposing parts of the dress to synthetic light from a 70 W 580 lumen halogen bulb for five weeks followed by three weeks of exposure to natural light through a plate-glass window without a UV filter.

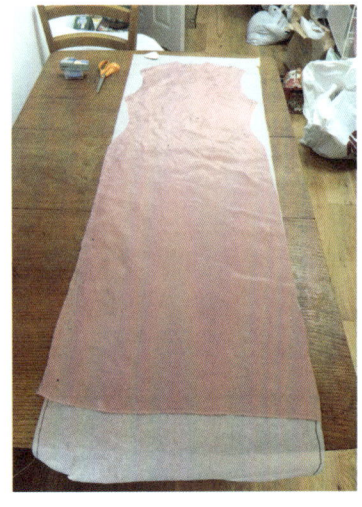

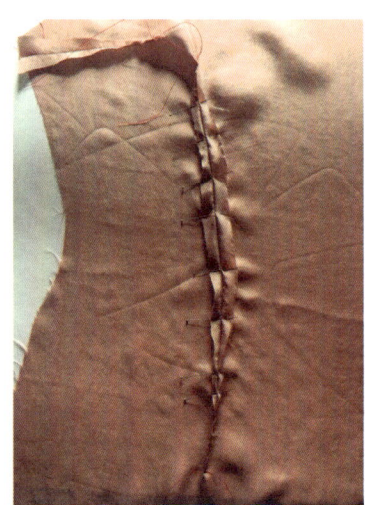

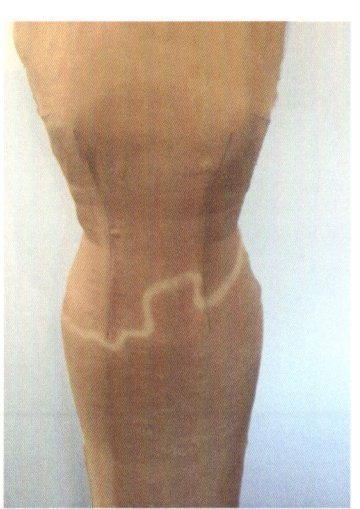

Oliver Sacks, neurologist

THE CASE OF THE COLOUR-BLIND PAINTER

AN EXCERPT

BRAIN BLINDNESS WAVELENGTH COLOUR SIGHT PERCEPTION

Early in March 1986 I received the following letter:

I am a rather successful artist just past 65 years of age. On January 2nd of this year I was driving my car and was hit by a small truck on the passenger side of my vehicle. When visiting the emergency room of a local hospital, I was told I had a concussion. While taking an eye examination, it was discovered that I was unable to distinguish letters or colours. The letters appeared to be Greek letters. My vision was such that everything appeared to me as viewing a black-and-white television screen. Within days, I could distinguish letters and my vision became that of an eagle – I can see a worm wriggling a block away. The sharpness of focus is absolutely incredible. BUT – I AM ABSOLUTELY COLOUR-BLIND. I have visited ophthalmologists who know noting about this colour-blind business. I have visited neurologists, to no avail. Under hypnosis I still can't distinguish colours. I have been involved in all kinds of tests. You name it. My brown dog is dark grey. Tomato juice is black. Colour TV is a hodge-podge …

Had I ever encountered such a problem before, the writer continued: could I explain what was happening to him – and could I help?

The weeks that followed were very difficult. "You might think," Mr I. said, "loss of colour vision, what's the big deal? Some of my friends said this, my wife sometimes thought this, but to me at least, it was awful, disgusting." He *knew* the colours of everything, with an extraordinary exactness (he could give not only the names but the numbers of colours as these are listed in a Pantone chart of hues he had used for many years). He could identify the green of Van Gogh's billiard-table unhesitatingly. He *knew* the colours in his favourite paintings, but could no longer see them, either when he looked or in his mind's eye. Perhaps he knew them, now only by verbal memory. It was not just that colours were missing, but that what he did see had a distasteful, "dirty" look, the whites glaring, yet discoloured and off-white, the black cavernous – everything wrong, unnatural, stained and impure.

Colour is not a trivial subject but one that has compelled, for hundreds of years, a passionate curiosity in the greatest artists, philosophers and natural scientists. The young Spinoza wrote his first treatise on the rainbow; the young Newton's most joyous discovery was the composition of white light; Goethe's great colour work, like Newton's, started with a prism; Schopenhauer, Young, Helmholtz and Maxwell, all in the last century, were all tantalized by the problem of colour; and Wittgenstein's last work was his *Remarks on Colour.* And yet most of us, most of the time, overlook its great mystery. Through such a case as Mr I.'s we can trace not only the underlying cerebral mechanisms or physiology but also the phenomenology of colour and depth of its resonance and meaning for the individual.

Mr I. also seemed to experience an excessive tonal contrast, with loss of delicate tonal gradations, especially in direct sunlight or harsh artificial light. But if the contrast was normal, or low, they might disappear from sight altogether. His despair of conveying what his world looked like, and the uselessness of the usual black-and-white analogies, finally drove him, some weeks later, to create an entire grey room, a grey universe, in his studio, in which tables, chairs, and an elaborate dinner ready for serving, were all painted in a range of greys. The effect of this, in three dimensions and in a different tonal scale from the "black and white" we are all accustomed to, was indeed macabre, and wholly unlike that of a black-and-white photograph. As Mr I. pointed out, we accept black-and-white photographs or films because they are *representations* of the world – images that we can look at, or away from, when we want. But black and white for him was a *reality*, all around him, 360 degrees, solid and three-dimensional, 24 hours a day. It was, he said, like living in a world "moulded in lead".

Subsequently, he said neither "grey" nor "leaden" could begin to convey what his world was actually like. It was not "grey" that he experienced, he said, but perceptual qualities for which ordinary experience, ordinary language, had no equivalent.

He was depressed once by a rainbow, which he saw only as a colourless semicircle in the sky. And he even felt his occasional migraines as "dull" – previously they had involved brilliantly coloured geometric hallucinations, but now even these were devoid of colour. He sometimes tried to evoke colour by pressing the globes of his eyes, but the flashes and patterns elicited were equally lacking in colour. He had often dreamed in vivid colour, especially when he dreamed of landscapes and paintings; now his dreams were washed-out and pale, or violent and contrasty, lacking both colour and delicate tonal gradations.

By the beginning of February, some of his agitation was calming down; he had started to accept, not merely intellectually, but at a deeper level too, that he indeed was totally colour-blind and might possibly remain so. His initial sense of helplessness started to give way to a sense of resolution – he would paint in black and white, if he could not paint in colour; he would try to live in a black-and-white world as fully as he could. This resolution was strengthened by a singular experience, about five weeks after his accident, as he was driving to the studio one morning. He saw the sunrise over the highway, the blazing reds all turned into black: "The sun rose like a bomb, like some enormous nuclear explosion," he said later. "Had anyone ever seen a sunrise in this way before?"

Inspired by the sunrise, he started painting again – he started, indeed, with a black-and-white painting that he called "Nuclear Sunrise", and then went on to the abstracts he favoured, but now painting in black and white only.

In these two months he produced dozens of paintings, marked by a singular style, a character he had never shown before. In many of these paintings there was an extraordinary shattered, kaleidoscopic surface, with abstract shapes suggestive of faces – averted, shadowed, sorrowing, raging – and dismembered body parts, faceted and held in frames and boxes. They had, compared with his previous work, a labyrinthine complexity, and an obsessed, haunted quality – they seemed to exhibit, in symbolic form, the predicament he was in.

Mr I. himself was actively curious about what was going on in his brain. Though he now lived wholly in a world of lightnesses and darknesses, he was very struck by how these changed in different illuminations; red objects, for instance, which normally appeared black to him, became lighter in the long rays of the evening sun, and this allowed him to infer their redness. This phenomenon was very marked if the quality of illuminations suddenly changed, as, for example, when a fluorescent light was turned on, which could cause an immediate change in the brightness of objects around the room. Mr I. commented that he now found himself in an inconstant world, a world whose lights and darks fluctuated with the wavelength of illumination, in striking contrast to the relative stability, the constancy, of the colour world he had previously known.

It had been shown in the 1960s that there were cells in the primary visual cortex of monkeys, in the area termed V1, that responded specifically to wavelength, but not to colour. And in the early 1970s it had been shown that there were other cells in the V4 areas that responded to colour but not to wavelength. These V4 cells, however, received impulses from the V1 cells, converging through an intermediate V2 structure. Thus each V4 cell received information regarding a large portion of the visual field. These two stages have an anatomical and physiological grounding: lightness records for each waveband being extracted by the wavelength-sensitive cells in V_1, but only being compared or correlated to generate colour in the colour-coding cells of V_4.

Mr I., it was clear, *could* discriminate wavelengths, but he could not go on from this to translate the discriminated wavelengths into colour. Mr I. was seeing with his cones, seeing with the wavelength-sensitive cells of V_1, but unable to use the higher order of V_4. For us, the output of V_1 is unimaginable, because it is never experienced as such and is immediately shunted on to a higher lever, where it is further processed to yield the perception of colour. Thus the raw output of V_1 never appears in awareness for us. But for Mr I. it did – his brain damage had made him privy to, indeed trapped him within, a strange, in-between state – the uncanny world of V_1 – a world of anomalous and, so to speak, prechromatic sensation, which could not be categorized as either coloured *or* colourless.

Mr I., with his heightened visual and aesthetic sensibilities, found these changes particularly intolerable. Colour perception had been an essential part of his creative identity, an essential part of the way he constructed his world – and now it was gone, not only in perception, but in imagination and memory as well. The resonances of this were very deep. At first he was intensely, furiously conscious of what he had lost.

But then, with the "apocalyptic" sunrise, and his painting of this, came the first hint of change, an impulse to construct the world anew, to construct his own sensibility and identity anew. Some of this was conscious and deliberate: retraining his eyes and hands to operate. But much occurred at a level of neural processing not directly accessible to consciousness or control. In this sense, he started to be redefined by what had happened to him – redefined physiologically, psychologically, aesthetically – and with this there came a transformation of values, so that the total otherness, the alienness of his new world, came to take on a strange fascination and beauty. "Gradually I'm becoming a night person," Mr I. said. He took to exploring other cities, other places, but only at night. He would drive to Boston or Baltimore, or to small towns and villages, arriving at dusk, and then wandering about the streets for half the night, occasionally going into little diners: "Everything in diners is different at night, at least if it has windows. The darkness comes into the place, and no amount of light can change it. They are transformed into night places. I love the nighttime," he said. "It's a different world: there's a lot of space – you're not hemmed in by streets, by people ... It's a whole new world."

Most interesting of all, the sense of profound loss, and the sense of unpleasantness and abnormality, so severe in the first months following his injury, seemed to disappear, or even reverse. A whole new world of seeing, of imagination, of sensibility, was born. In terms of his painting, after a year or more of experimenting and uncertainty, Mr I. has moved into a strong and productive phase, as strong and productive as anything in his long artistic career. His black-and-white paintings are highly successful, and people comment on his creative renewal. Although Mr I. does not deny his loss, and at some level still mourns it, he has come to feel that his vision became "refined", "privileged", that he sees a world of pure form, which the rest of us are insensitive to.

Arturo Fuentes, composer

SNOWSTORM

DYNAMIC   ELUSIVE   CONTRAST

William Turner's *Snowstorm* (1842) invokes a dark and nebulous tone; qualities that I wanted to reconstruct in my composition. Thrown forcibly around a vortex, the colours in the painting mix indefinitely. Lightness, brightness, motion and depth also unravel the canvas. In musical terms, these aspects of Turner's painting made me rethink the articulation of sound in my music. I created a nebulous soundspace, projected vigorously around a vortex with several instrumental layers. In the composition, as in Turner's painting, delicate shapes emerge from a chaos of colours and sound. This image of light that I discovered in the English painting, I shaped and reshaped in my music.

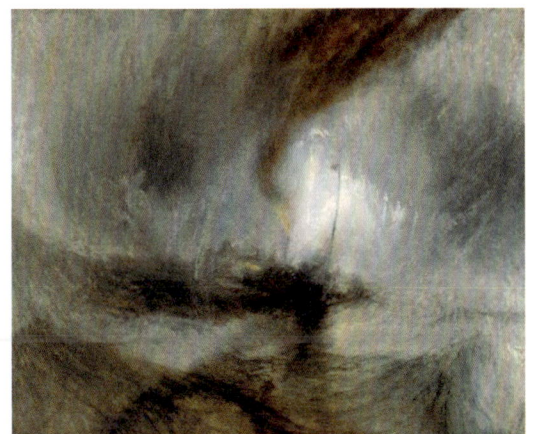

My *Snowstorm* was commissioned by the Ensemble Intercontemporain. It is a piece for 20 musicians, of 20 minutes duration. Premiere 11 February 2015, Opéra de Bordaeaux. Conductor: Matthias Pintscher.

Grace Note
RAIN + B
electrónica
se mantiene
durante
1a. parte.
pare.

Raihana Ferdous, geographer

CHEAP ENERGY FOR ALL?

ENERGY   OFF-GRID   SUN   POLITICS

In April 1954 the *New York Times* published the breaking news that sunlight could be converted into energy, claiming, "the beginning of a new era – the harnessing of the almost limitless energy of the sun for the uses of civilization".

Since then solar energy has emerged as a leading alternative to fossil-fuel-generated light. In 2013 it was reported that the global photovoltaic solar energy market was generating at least 139 GW electricity, that is, 1% of total worldwide electricity consumption.[1] The United Nations has a goal of sustainable energy for all by 2030, and solar energy programmes are among the most popular global initiatives to achieve this. It encourages a multi-stakeholder approach that brings

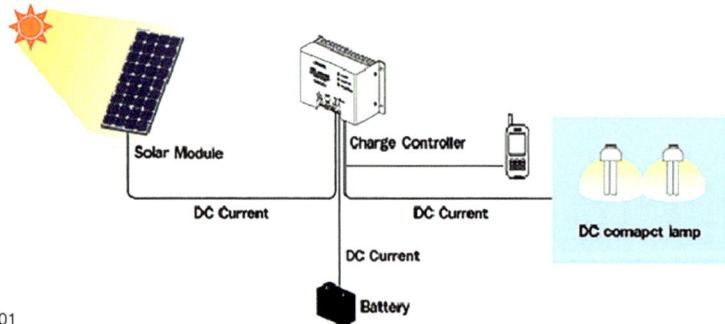

01

together governments, businesses and local communities in bilateral partnerships. My ethnographic research focuses on bi-lateral partnerships in Bangladesh, assessing its effectiveness and socio-economic implication in often poor, off-grid areas, and what we can learn moving forward.

SOLAR ENERGY PROGRAMME IN BANGLADESH

Bangladesh is a small but densely populated country in South Asia. With a population of approximately 156 million people, Bangladesh is situated in the global latitudes 20° and 27°N, and longitudes 88° and 93°E, which can be considered an ideal location for

01   A Solar Home System (SHS) comes with four essential components: a solar panel (photovoltaic module), a charge controller, a battery and some connections for light bulbs, and a mobile phone charging connection. The system operates at a rated voltage of 12 and provides power for small electrical loads such as fluorescent luminaries, radios, cassette players, and small black-and-white television sets or similar low-power appliances for about four hours a day. A typical SHS generally comprises 2 to 8 DC lights ranging from 6 W to 20 W, depending on the capacity of the module.

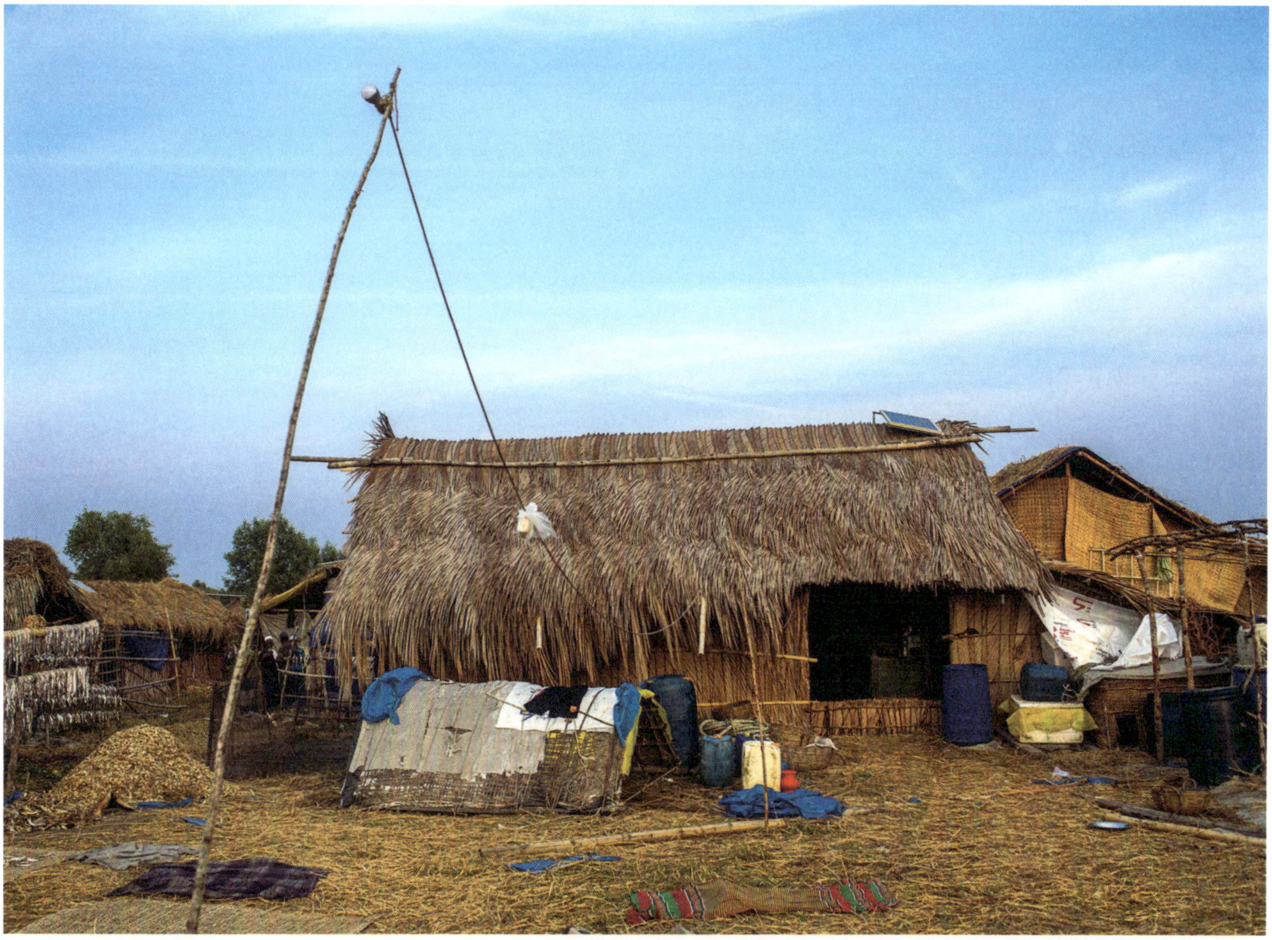

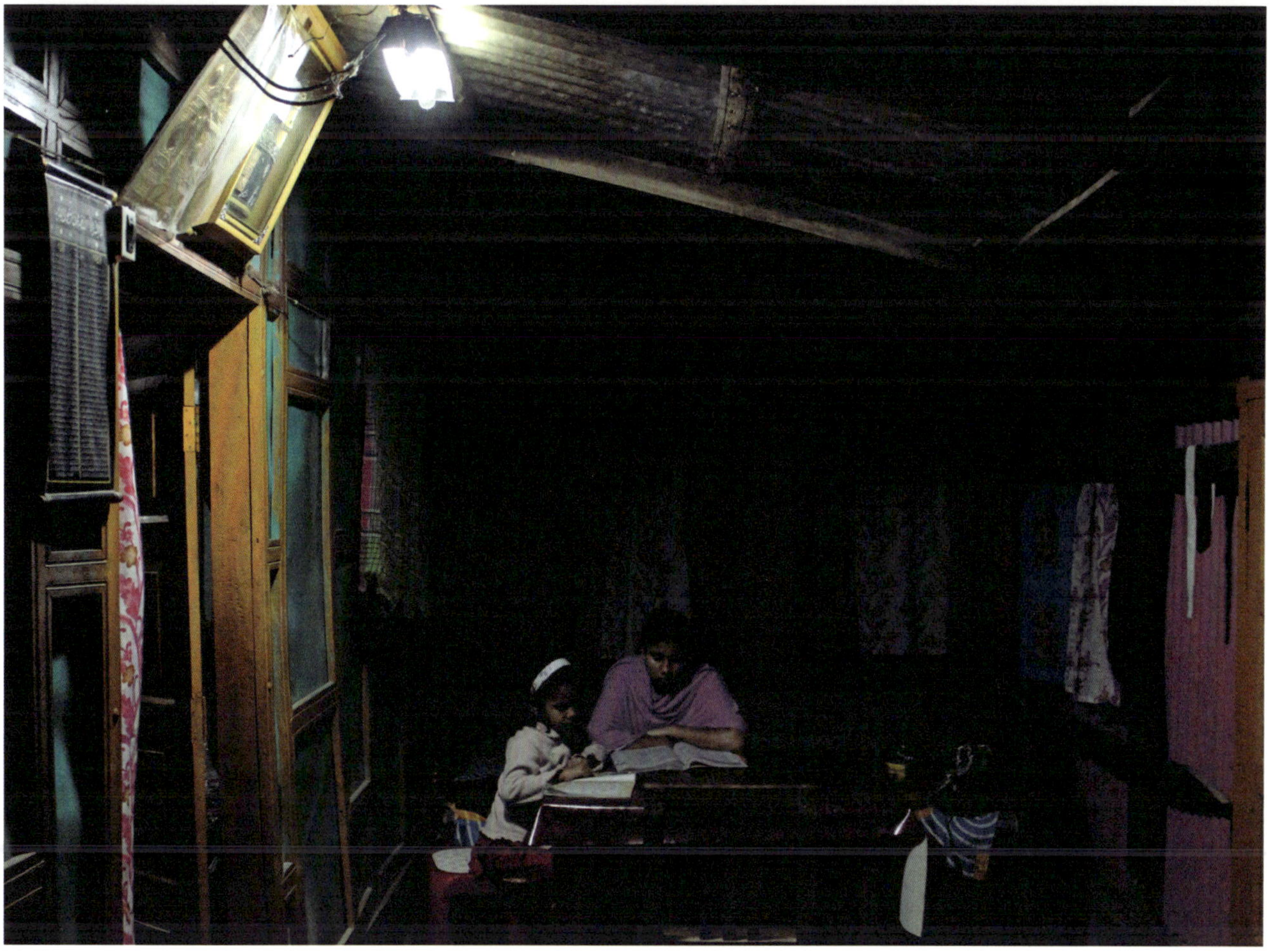

02–03

solar energy reception. The energy infrastructure of this country is reasonably small and poor, one of the lowest per capita energy consumptions, in the world. Only 62% of the total population has access to electricity, with a per capita availability of 321 kWh per annum. The Bangladeshi government is striving to replace kerosene lamps, which are toxic, environmentally unfriendly and dangerous to use, and has been promoting the Solar Home System programme, with the help of international donor agencies. It is a market-based approach, where IDCOL (Infrastructure Development Company Limited), a state-owned financial bank, lends money to the partner organizations (POs), ie, local NGOs and private organizations. To date 13 million beneficiaries are receiving solar electricity in this way, ie, around 9% of the total population are accessing light through solar energy. All this might lead one to think that the introduction of solar energy in Bangladesh is fast turning into a success story. My research exposes three major challenges of the current approach.

The economic mechanism in which it operates makes a Solar Home System expensive to the end user. To purchase a Solar Home System, low-income families have to borrow from the NGOs, and at 12% interest these families monthly spend a similar amount in repayments as they would on kerosene. It typically takes a family three years to repay the loan plus interest. But that's not the end: every four to five years they will need to change the solar battery, at a cost of one-third of the price of a complete Solar Home System. This means solar energy is still expensive for a large proportion of the population [02]. Furthermore, because the systems are so expensive solar users end up paying four to eight times more for a solar electricity unit than they would if they had access to grid electricity.[2] Inevitably, solar energy remains beyond the reach of many Bangladeshis.

A Solar Home System only has the capacity to produce a limited amount of electricity for small and selected electronic appliances. A typical SHS comes with a four-hour back-up battery [01]. These limited hours of electricity supply make it difficult for any household to cope with every day use as can be seen in [03]. As a solar user explained: "You always need to plan your work,

04   This tea stall is situated next to a busy road, but you can hardly see the road or anything else, because there are no streetlights. The Solar Home System programme in Bangladesh may have brought light inside, but outside it is still dark. The market-led Solar Home System programme allows those individuals to light up their houses, but it has no provision for communal space. Roads and streets are still in the dark; women do not feel comfortable going out after dark.

04

prioritize your needs and justify your use, otherwise, you will be sitting in the dark just when you need light the most." This is a real challenge for solar users since they need to continually prioritize their needs and adjust their life accordingly.

Currently POs sell the Solar Home Systems to families and small businesses. There is little to no implementation of solar energy in public spaces or infrastructure. The tea stall in [04] is situated next to a busy road, but you can hardly see the road or anything else, because there are no streetlights. The Solar Home System programme in Bangladesh may have brought light inside, but outside it is still dark. The market-led Solar Home System programme allows those individuals to light up their houses, but it has no provision for communal space. Roads and streets are still in the dark; women in particular do not feel comfortable going out after dark.

If we are to achieve the UN global target of sustainable energy for all in Bangladesh, and solar energy is to be a significant player in this field, the above issues need to be resolved, starting with making this technology truly affordable and worthwhile for all.

1   http://www.nrel.gov/education/
2   http://idcol.org/home/solar pdfs/educational_resources/
    high_school/solar_cell_history.pdf
3   I. Sharif, M. Mithila (2013), "Rural Electrification using
    PV: the Success Story of Bangladesh", *Energy Procedia*,
    33:343–354
4   http://www.ren21.net/Portals/0/documents/Resources/
    GSR/2014/GSR2014_full%20report_low%20res.pdf
5   http://www.rehab-bd.org/An_Evaluation_of_3percent_
    Roof_Top_Solar_Panel_Policy_in_Bangladesh.pdf

01 220 v clear glass bulb with pip. Two looped carbon filaments with supporting brackets to separate the filaments, fixed with grains of brown sugar. Bayonet fitting, intact. Circa 1900–1910.

03 220 v bulb with mirror glass and pip. Two carbonized yarn filaments with supporting brackets, intact, possibly by Mr Swan. Circa 1910.

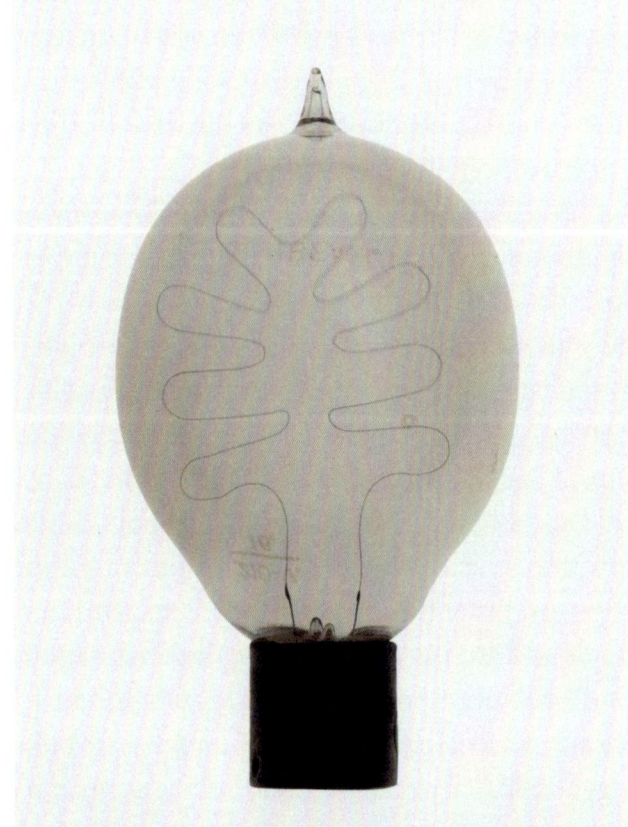

02 "Sunbeam" 210 v 16 cp lamp, clear glass bulb with pip. Very floppy single carbon filament arranged in wavy pattern with 8 "fingers". Platinum lead-in wires. Cement retaining pockets in glass envelope anchoring bulb to cap. Circa 1897.

04 An Ediswan 100 v 16 cp heavy ruby-red glass with pip lamp, bayonet fitting with plaster of Paris, single coil, intact, carbon filament.

THE BILL CARLTON COLLECTION OF INCANDESCENT LAMPS

LIGHT BULB   CARBON   INNOVATION

This rare and unusual collection comprises approximately 500 bulbs, including early carbon lamps with loop contacts, tantalum lamps and ancient lamps, spanning principally 1890–1920.

In 1802 Penzance's most famous son, Humphry Davy, passed an electric current through a strip of platinum and, in so doing, created the first incandescent light. In 1809 he connected two wires to a battery and attached a charcoal strip between the other ends of the wires. The charged carbon glowed, making the first arc lamp.

These experiments were the precursors to the scores of experiments that 75 years later led to the electric light that we are now so familiar with today.

This collection gives us a fascinating glimpse into an emerging technology of the 1800s and some of the earliest attempts to mass-produce it. A technology that completely revolutionized our lives and one that today, we take entirely for granted.

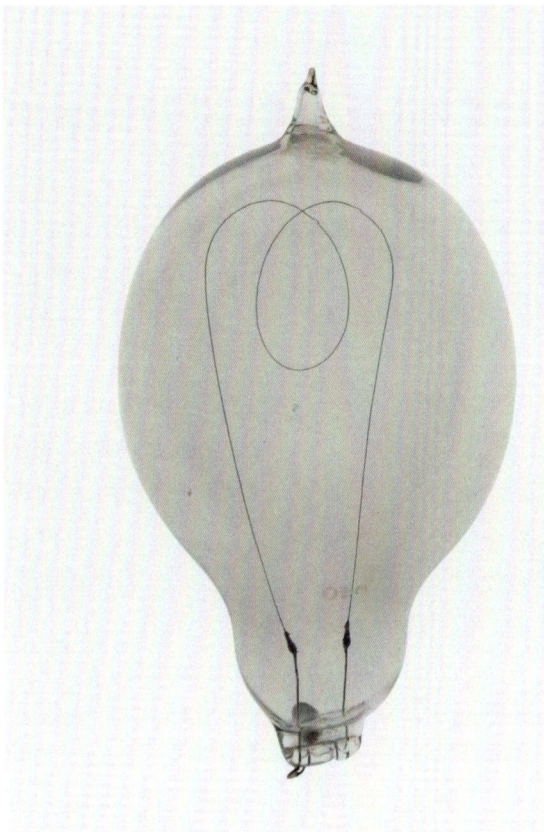

05   Clear glass lamp, marked 940, platinum lead-in wires with loop protruding through glass. Single carbon filament, with one coil.

07   "Elblight Sunbeam" lamp. Lemon-shaped bulb with pip glass, exterior sprayed with blue lacquer. Single coil, single carbon filament. Unusual ELB double pin fitting, circa 1900.

06   Ediswan 110–8 A29, lemon-shaped bulb with pip. Single carbon filament with single coil. Platinum lead-in wires. Straight-sided brass bayonet fitting with single centre contact in plaster of Paris. Circa 1890s.

08   STEARN 230 v 8 cp. Clear glass lamp with pip. Long platinum lead-in wires, double carbon filaments. Bayonet fitting in brass and black moulded vitriolite, ground top. Circa 1883–90.

09   Early twisted, ribbed Ediswan candle lamp 110–8. A, with single carbon filament and bayonet fitting with plaster of Paris. Circa 1897.

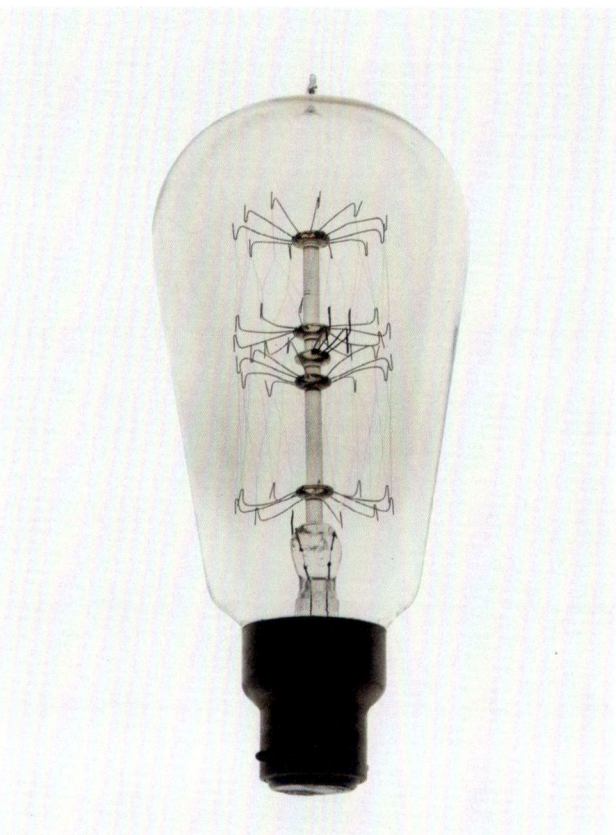

11   German manufacture Tantalum lamp 230–25 xlk, Platinum lead-in wires, drawn tantalum wire filament, woven over 4 levels of copper supports, 10 arms on each level. Earliest examples of this bulb date from 1906.

10   A heavy moulded glass lamp in Bristol blue branded: GABRIEL & ANGENAULT DEPOSE. Bayonet fitting with black vitriolite.

12   Unusually shaped stepped, ribbed, clear lamp. Double coiled, double carbon filament. Bayonet fitting with steatite. Circa 1902.

13　ELB system, frosted globe with pip lamp, multiple coiled carbon filament, glazed white porcelain cap with double pin fitting. (ELB: "Electric Lighting Boards") Circa 1900.

14　Nerst Lamp. Extremely frosted sphere with Beretta housing.

Kobus Kuipers, Anouk de Hoogh, Nir Rotenberg and Boris le Feber, nano-scientists

CONTROLLING AND EXPLOITING LIGHT'S SMALLEST FEATURES

NANOPLASMONICS   SCALE   METAMATERIALS   CONTROL   QUANTUM OPTICS
DIALECTRIC   INFORMATION CARRIER

INTRODUCTION

Light is crucial to humans' arguably most important sense: vision. The importance of light in today's society, however, goes well beyond seeing things; it connects the world through the Internet, it allows massive amounts of data to be stored and read, it forms a route to sustainable energy, and it allows minute quantities of material to be detected. Light is an electromagnetic wave composed of oscillating electric and magnetic fields, the one never occurring without the other, their oscillations intertwined through James Clerk Maxwell's equations on electromagnetism. The wave nature of light combined with the optical properties of naturally occurring materials makes it hard to control it at length scales (much) smaller than its wavelength. Nevertheless, such control – the central goal of nanophotonics – presents a very rewarding challenge since it enables both cutting-edge science and potential application in modern technologies.

In the last decade, the fields of nanoplasmonics and photonic crystals have opened up the nanoscale for optical control.[1-3] In nanoplasmonics, the optical properties of tiny metallic nanostructures are tuned through their geometry, enabling a huge control over nanoscale light. In photonic crystal structures this optical control is achieved through interference and multiple scattering in periodic dielectric structures. Both the flow and emission of light can be controlled at these small-length scales, giving rise to new science and novel applications. In photonic crystal cavities light has been confined to volumes smaller than the cubed wavelength while maintaining a high storage time of the light, ie, a high quality factor Q,[4, 5] thus enabling quantum optics at the nanoscale.[6-8] Photonic crystals have been used to slow down light and enhance nonlinear optics.[9, 10] Plasmonic nanostructures can confine light to deep subwavelength volumes,[11] can guide light along wires more than an order of magnitude smaller than the optical wavelength,[12] can transport quantum information[13] and enable ultrafast control of light propagation.[14] Metal structures have even been used to create metamaterials to gain unprecedented control over, amongst other things, the refractive index, by exploiting both the electric and magnetic vector components of light.[15-21] These days the confinement of the light has become so extreme that fundamental limits have been reached: on the smallest-length scales a non-local response gains in importance and limits the optical behaviour[22] and optically induced electron tunnelling has been observed to affect the optical response of plasmonic nanostructures.[23]

Interestingly, freely propagating light beams can already contain features that are significantly smaller than a wavelength. In fact, these features have a size that puts nanophotonics to shame: they are singularities, ie, they have size zero. While in normal light beams, the amplitude, phase and polarization can be thought to vary smoothly in space, in the presence of singularities this is no longer the case: *light is said to have acquired structure*.[24, 25] For example, the beautiful optical patterns on the bottom of swimming pools, so-called caustics, exhibit a macroscopic structure that can readily be observed by eye. However, the various vector components of the light actually also conspire to form a network of polarization singularities that can, of course, not be directly observed by the eye. Structured light may acquire angular orbital momentum,[26] exhibit knots[27, 28] and survive highly nonlinear behaviour,[29] all in free-space light beams.

It turns out that optical eigenmodes of photonic nanostructures also contain optical singularities. For example, a so-called W1 photonic crystal waveguide (PhCWG), a waveguide formed

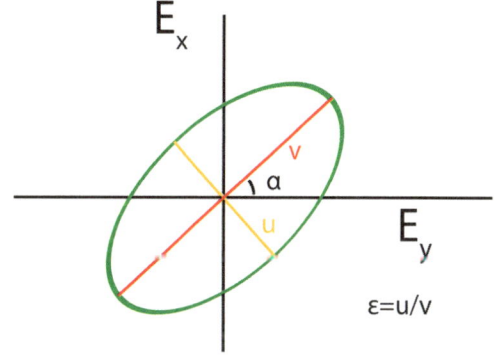

01   Polarization ellipse for the electric field in the x, y plane. The ellipticity is given by the ratio of its long and short axes: ε=u/v. Its orientation angle is given by α.

by removing one line of holes in a two-dimensional photonic crystal [04], contains four polarization singularities in the electric[30] and magnetic[31] fields per unit cell.

POLARIZATION SINGULARITIES

The electric (or magnetic) field of light in a single plane can oscillate in two (orthogonal) directions, for example, x and y. At any point in the plane the vector will describe an ellipse as a function of time [01]. The shape of the ellipse is determined by the magnitude and relative phase of the electric field in the x- and y-direction. The ellipse can be described by two parameters. In the first place the "ellipticity" ε is the ratio of the short to the long axis of the ellipse (u/v) and is represented by a number between −1 and +1, with the sign denoting the handedness of the light. In the extremities of ε, $\varepsilon = -1$ or $\varepsilon = +1$, the light is circularly polarized. On the other hand, for $\varepsilon = 0$ the light is linearly polarized. In the second place α represents the orientation of the ellipse with respect to a chosen axis.

Positions where one of the parameters describing the polarization ellipse becomes degenerate or "singular" are particularly interesting. Points in light fields where this occurs are called polarization singularities. One example of a polarization singularity is a location where the handedness of the light field is undefined/degenerate, which is the case for linear polarization. In a plane this typically occurs along lines, so-called L-lines. Another example is α being undefined (circular polarization). In a plane this occurs in so-called C-points. Please note that these singularities are infinitesimally small.

The study of optical singularities is not just of fundamental interest, as these structured light fields can also be of practical use. Recently, our group has begun exploring how the light fields near optical singularities can be exploited to control nanoscale emitters. Specifically, we want to know whether we can control the direction in which such an emitter sends photons, when it emits near the location of a particular type of singularity. If this directionality could be controlled, then our structures in which the singularities occur could act as a novel interface between the state of a quantum emitter – for example a quantum dot or vacancy centre in a nanodiamond – and photons.

POLARIZATION SINGULARITIES NEAR A SUBWAVELENGTH HOLE

Arguably the most simple nanophotonic structure is a subwavelength hole. Interestingly, when a surface plasmon polariton (SPP) beam scatters from a subwavelength hole in a metal film, the interference between the incident SPP beam and the scattered light field leads to a plethora of optical singularities in the near field.

The bottom panel of [02] schematically depicts the aforementioned situation: an incident Gaussian SPP beam (width s), travelling along x impinges on a hole (diameter d) and causes a seemingly circular scattered field. Please note that the actual pattern of the scattered field is not circular due to a subtle interference of the electric and magnetic response of the hole.[32] The total resulting light field is the result of the interference of the incident and the scattered field. The top panel of [02] depicts the calculated amplitude of the resulting electric field. A clear, roughly parabolic pattern of ripples is visible, of which the period along x in front of the hole is half the wavelength of the SPPs.

From the calculations we can reconstruct the polarization ellipse in the x, y plane for every point. Figure [03] displays a false-colour map of the orientation α of the ellipse as a function of position. It is immediately obvious that all orientations of the ellipse are present in the vicinity of the hole. To investigate whether the light field also contains C-points where α is undefined, we use contour lines that connect points of equal α (black solid lines). When these lines come together in a single point, α has to be undefined, as it cannot have all these orientations at the same time. Around the C-point all orientations of the ellipse between 0 and π occur. Depending on the sequence in which these values occur when a circle is traced around the C-point (in the counter clockwise direction), a topological charge of either +½ or −½ is assigned to the singularity. We observed two strings of singularities on either side of the hole along the direction of the incident SPP beam, of which the inner and outer string have an opposite topological charge.

For a given situation the pattern of singularities is fixed in space. However, by subtly varying one of the parameters of the experiment the positions of the singularities can be made to change. For example, in this system the total electric field can be manipulated by varying the ratio between the incident and scattered field amplitudes. This ratio can be controlled by changing the hole diameter and thereby how strong it scatters the incident wave. The ratio between the respective field amplitudes determines the distance of the strings of polarization singularities to the hole. Interestingly, the optical amplitude at the location of the C-points will change accordingly: the closer to the hole, the

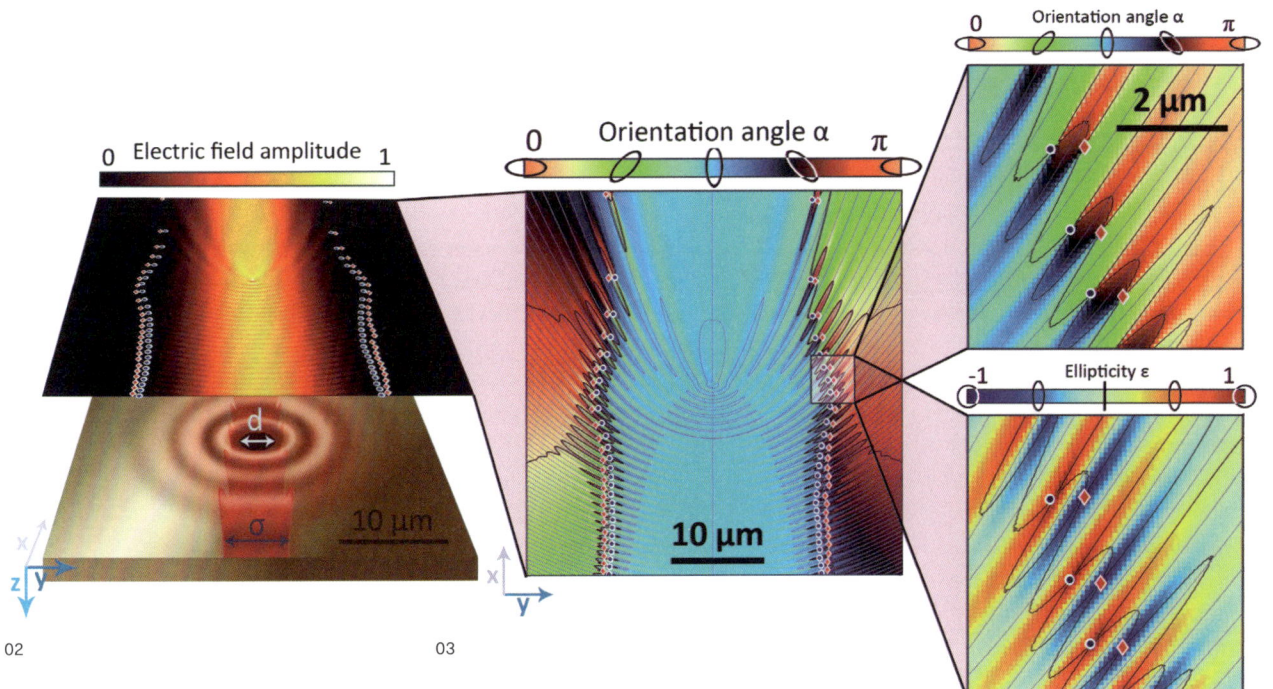

02   Electric field amplitude 0 — 1

0   Orientation angle α   π

0   Orientation angle α   π

2 µm

-1   Ellipticity ε   1

10 µm

10 µm

02   03

02   Schematic depiction of the magnitude of the electric field around a single subwavelength hole in a thick metal film that is illuminated by a Gaussian SPP beam (top). The normalized amplitude in the x, y plane 10 nm above the surface shows that interference occurs between the incident and the scattered SPP fields.

03   A false-colour representation of the orientations of the polarization ellipse of the in-plane electric field. The subwavelength hole is situated (not visible) at the centre of the map. A zoom-in of the shaded area presents both the orientation (top) and the ellipicity (bottom) of the fields. C-points are marked with blue circles and red diamonds mark C-points of opposite topological charge. The thin black lines are L-lines where the "polarization" is linear.

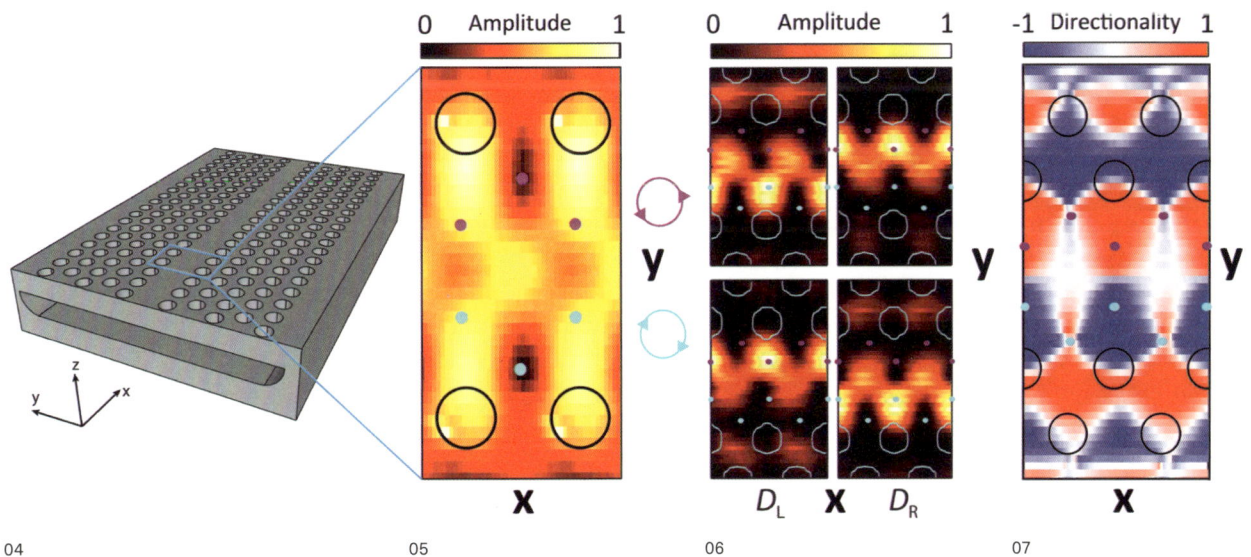

0   Amplitude   1

0   Amplitude   1

-1   Directionality   1

D_L x D_R

04   05   06   07

04   Schematic depiction of a photonic crystal waveguide. The waveguide is formed by drilling a row of holes in a periodic arrangement of holes in a 220-nm-thick Si (silicon) membrane.
05   Normalized amplitude of the x-component of the electric field measured roughly 10 nm above the waveguide with a polarization-sensitive near-field microscope. The blue and red dots denote the locations of polarization singularities. The solid black lines denote the holes of the surrounding photonic crystal.

06   Measured amplitude of the light detected by the detectors on the right and left of the structure when a left-handed (top) or a right-handed (bottom) circular dipole emits into the waveguide.
07   The directionality of the emission into the waveguide as determined from the measurements represented in [06].

higher the amplitude. It turns out that by varying the size of the hole, the amplitude of the C-points can be varied over three orders of magnitude.[33] This is important for possible applications of the singularities.

EXPLOITING LOCAL HELICITY FOR EMISSION CONTROL

Another ubiquitous nanophotonic structure, the photonic crystal waveguide (PhCWG) [04], has eigenmodes that are very rich in structure.[30, 34] Figure [05] depicts the local amplitude in two unit cells measured with a near-field microscope.[35] It is clear that the amplitude is modulated with the same periodicity as that of the surrounding photonic crystal (the solid black circles represent the holes of the crystal). The dots in the figure represent locations of C-points found with the same method as that described in the previous section. The C-points on opposite sides of the waveguide (red vs blue dots) have an opposite-handedness. More interestingly, the handedness also flips when eigenmodes are considered with opposite propagation direction (left-to-right vs right-to-left). Consequently, when we place a circular dipole source near these locations, the direction in which it emits should be determined by the handedness of the local light fields.

We use the emission of a subwavelength nanoprobe to test this hypothesis. Our nanoprobe consists of a perfectly circular subwavelength aperture consisting of a glass core surrounded by aluminium (Al). The aperture is connected to a single-mode fibre. By controlling the polarization state coupled to the other end of the fibre we can make the aperture act like a linear or circular dipole. This dipole is then scanned across the PhCWG structure and two detectors at either end (left or right) collect the light coupled to the waveguide. Their signals are represented as D_L and D_R, respectively. The amount of energy radiated left- and rightwards into the waveguide by *linear* dipoles is found to be identical (symmetric) for all positions and orientations of the dipole near the waveguide (data not shown).

Strikingly, for circular dipole emission the measured left and right emission maps are drastically different [06]. For a given handedness of the dipole the excitation patterns as measured by detectors D_L and D_R are clearly different. In fact, when the measured amplitude of one detector is highest, the amplitude on the other detector is (close to) zero. Hence, this measurement immediately suggests that the left-right emission symmetry can be broken. Furthermore, we observe that this symmetry breaking is reversed with a change of dipole helicity.[36]

To quantify the emission directionality and to test if the emission direction is determined by dipole helicity we define:

$$\Delta = \frac{1}{2} \frac{D_L^{LCP} - D_R^{LCP}}{D_L^{LCP} + D_R^{LCP}} - \frac{D_L^{RCP} - D_R^{RCP}}{D_L^{RCP} + D_R^{RCP}}$$

where Δ is the emission directionality and D_L and D_R are measured left- and rightwards emitted intensities for right- (RCP) and left-handed (LCP) circularly polarized dipole. If $\Delta = \pm 1$ a circular dipole emits perfectly directionally into the waveguide depending on its helicity. The experimentally obtained Δ maps reveal near unity directionality when the tip is placed on a polarization singularity [07]. Furthermore, we observe that large regions of high Δ are available away from the singularities (dark-red and blue regions).

Although we use a classical dipole source to map the directionality of the emission, the same circular transition dipole orientations are associated with the decay of quantum emitters between different spin-states. That is, our measurements show that, by using the light fields near optical singularities, we are taking an important first step towards an on-chip platform capable of interfacing the spin of solid-state emitters and directionality of photons. Such a platform could, in principle, be used to entangle emitter spin and photon directionality, and even be used to create quantum architecture.[37, 38]

CONCLUSIONS

We have shown that polarization singularities can occur near nanophotonic structures. The singularities can be manipulated by passively or actively varying parameters of the system (geometry or frequency). In addition to being of fundamental interest, our demonstration of dipole helicity to photonic path coupling may also impact many applications. A particularly appealing example is the interface between the spin of an electron in a solid-state emitter and the propagation direction of light. Solid-state emitter spin is often proposed to store quantum information, whilst information encoded in the path of light can be used for the transport of information and for basic quantum arithmetic. The transition to and from different spin states is associated with circular dipoles of opposite helicity. Because we observe that these dipoles radiate directionally when placed on polarization singularities in PhCWs, polarization singularities could enable an on-chip interface that forms a novel route towards on-chip quantum information processing.

REFERENCES

1 W.L. Barnes, A. Dereux, T.W. Ebbesen (2003), *Nature,* 424:824
2 A. Polman (2008), *Science,* 322:868
3 J. D. Joannopoulos, S.G. Johnson, J.N. Winn, R.D. Meade (2008), *Photonic crystals: Molding the flow of light* (2nd edition (Princeton University Press)
4 B.S. Song, et al (2005), *Nature Materials,* 4:207
5 T. Tanabe, et al (2007), *Nature Photonics,* 1:49
6 I. Fushman, et al (2008), *Science,* 320:769
7 L. Sapienza, et al (2010), *Science,* 327:1352
8 Y. Sato, et al (2012), *Nature Photonics,* 6:56
9 B. Corcoran, et al (2009), *Nature Photonics,* 3:206
10 P. Colman, et al (2010), *Nature Photonics,* 4:862
11 D.K. Gramotnev et al, (2010), *Nature Photonics,* 4:83
12 E. Verhagen et al (2009), *Phys Rev Lett,* 102:203904
13 A.V. Akimov et al (2007), *Nature,* 450:402
14 K.F. MacDonald (2009), *Nature Photonics,* 3:55; M. Aeschlimann, et al (2007), *Nature,* 446:301
15 J.B. Pendry, et al (1999), *IEEE Trans Microwave Theory Tech,* 47:2075
16 R.A. Shelby, et al (2001), *Science,* 292:77
17 C. Enkrich, et al (2005), *Phys Rev Lett,* 95:203901
18 G. Dolling, et al (2006), *Science,* 312:892
19 C.M. Soukoulis, et al (2007), *Science,* 315:47
20 J. Valentine, et al (2008), *Nature,* 455:376
21 N. Liu, et al (2009), *Nature Photon,* 3:157
22 C. Ciraci, et al (2012), *Science,* 337:1072–1074
23 P. Cristofolini, et al (2012), *Science,* 336:704
24 J.F. Nye (1999), *Natural focusing and fine structure of light: caustics and wave dislocations,* Taylor & Francis
25 For a recent review: M.R. Dennis, et al (2009), *Prog Optics,* 53:293

26 For a recent review: S. Franke-Arnold, et al (2009), *Laser & Photon Rev,* 2:299
27 W.T.M. Irvine, et al (2008), *Nature Physics,* 4:716
28 M.R. Dennis, et al (2010), *Nature Physics,* 6:118
29 M. Zürch, et al (2012), *Nature Physics,* 8:743
30 M. Burresi, et al (2009), *Phys Rev Lett,* 102:033902
31 N. Rotenberg, et al (2015), *Optica,* 2:540
32 N. Rotenberg, et al (2012), *Phys Rev Lett,* 108:127402
33 A. de Hoogh, et al (2014), *J Opt,* 16:114004 1/8
34 R.J.P. Engelen, et al (2009), *Phys Rev Lett,* 102:023902
35 N. Rotenberg and L. Kuipers (2014), *Nature Photonics,* 8:919–926
36 B. le Feber, et al (2015), *Nature Communications,* 6:6695
37 I.N. Söllner, et al (2015), preprint at http://arxiv.org/abs/1406.4295
38 A.B. Young, et al (2014), preprint at http://arxiv.org/abs/1406.0714

The work described here was carried out by the NanoOptics Group in the Center for Nanophotonics at the FOM-institute AMOLF (Amsterdam, NL). This work is part of the research programme of the Foundation for Fundamental Research on Matter (FOM), which is part of the Netherlands Organization for Scientific Research (NWO).

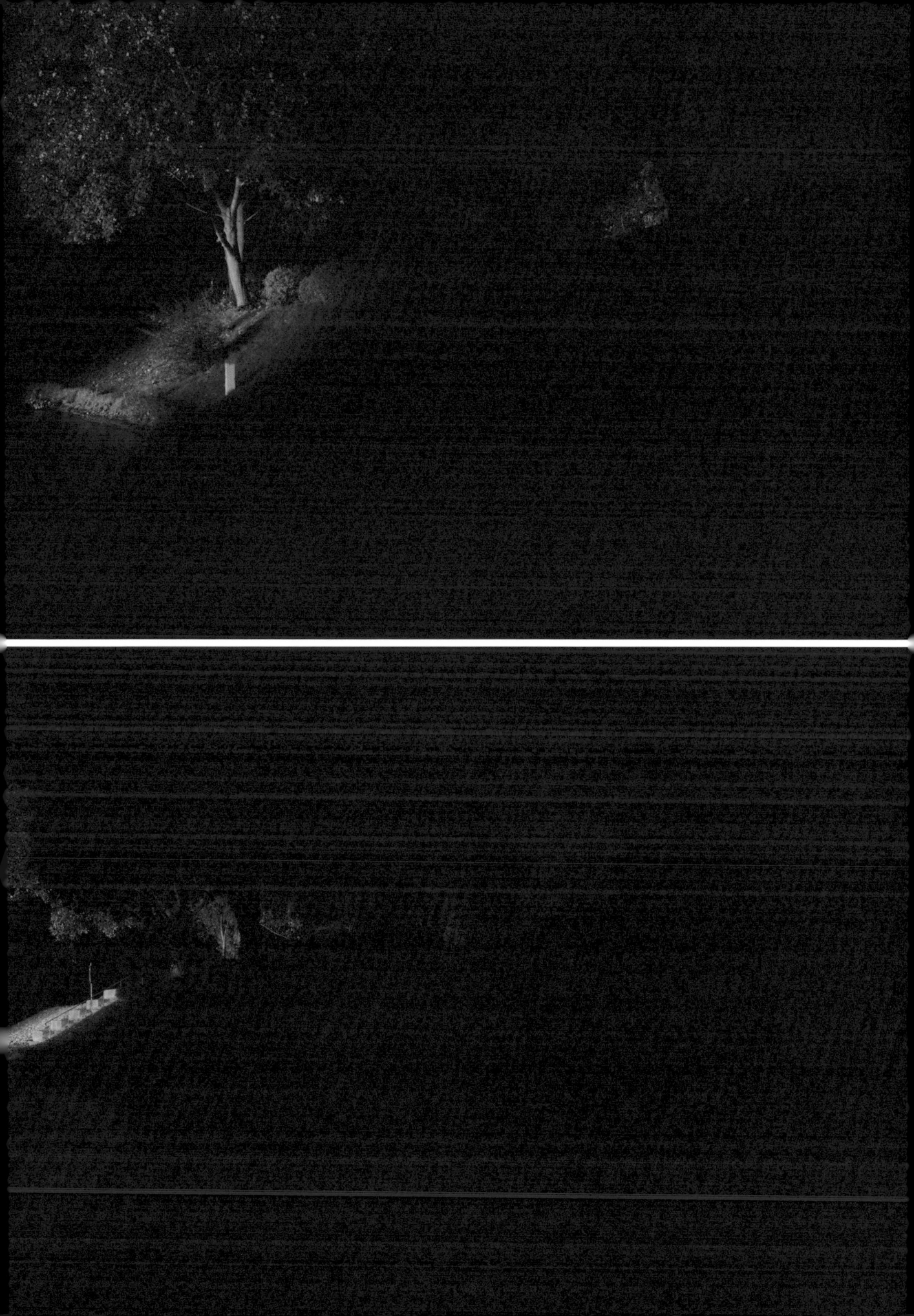

ZOOM

PHOTONS WAVE REFLECTION CHANCE IMAGINATION

imagine you and me
walking down a principal street say
    somewhere here in Amsterdam
and there's a store window
and someone in the store window is working
    with a mannequin
... a shop-window dummy

and the two of us ... we ... see the man in the
    store window
and he sees us, brightly illuminated
but we see something else
we see an image of ourselves in the window
dim, but it's there
it's enough for us to see: that's us
that's you, that's me

nothing as light as light
and nevertheless we find ourselves trapped
    by it
turned to stone if you will in the window ...
locked for a moment in the *oneness* that
this reflection-image is making of the *two*
    of us that we are just as well
at this very moment, this instant in time

as if light—1–2 with time—is playing a game
    with us
anticipating inconsiderately the breathless
    Abyss
the unfathomable depth of our *actual*
    becoming one, later
the wrestling panting, enslaved in each other's
    arms, legs, lips
alternately conquered, alternately conqueror
beyond control
abundant moment

so, imagine the two of us and light,
    sunlight, photons in this case,
particles of light, that illuminate the
    tip of our nose
and then head for the store window

sometimes a particle goes through
because the man in the window
    sees us
and sometimes it's reflected because
    we see ourselves (at the same time)
dimly but unmistakably the two
    of us
length, age, colour, sex,
flavour, mass, charge, spin—
what we're wearing while we're still
    dressed

now when are they reflected?
and when do they go through?

it turns out about 96% of the photons
    go through the window
and 4% reflect

and so we ask ourselves are these
    reflected photons
any different from the ones that went
    through?

and the answer, surprisingly (to us) is 'no' ...
   they're all the same

now what determines whether a photon will
   go through the window
or be reflected

and guess what: it's a random process

but Nature doesn't plays dice, does she?
No ... or yes ... what ... well no ... yes ... or
does she no yes or ...

but it's exactly what she does
she says ok here goes a photon, rolls a dice
if it's anywhere from 1 to 96 the photon
   goes through
but if it's from 97 to a 100, the photon is
   reflected

that's Nature as she is Absurd

and *"If you think you understand it, you don't*
   understand it."

imagine
you and me again standing in front of this
   store-window
somewhere here in Amsterdam

in front of the image of ourselves reflected
   in the window
while at the same time someone at our
   left can see
someone who's standing at our right
which means that the light is going right
   across this way
waves/particles going this way
between the person to our left and to our right
and particles waves going that way
in between the store-window dresser
who's dressing the store-window dummy …
   and us
this way, that way,
particles which are at the same time to
   be regarded as waves
travelling in each and every possible way
a complete network
something shaky
particles & waves, waves & particles
this way, that way
particles & waves, waves & particles
not one thing

like
two i's fighting for the same dot
which is the title of a painting by René Daniëls
two i's fighting for the same dot
both the stroke of the brush and the dot

that characterize the art of painting

like the way in which an insight, a thought, an
    idea
that happens, that comes, in a particular
    instant in time
at the same time needs to be defined as being
    spread out,
stretched over the words, the minutes,
    hours, years
of thinking that's compressed, squeezed,
    united in the thought,
in the very moment when it comes

particle & wave, wave & particle

this being in two minds
and getting lost
together with the 4% of photons
that *do* reflect at, in, on, against, upon the
    shop window
for reasons that will not become more
    reasonable when we try to explain them
finally being absorbed by, into our irises ...
brought to a stand-still, I guess, in the black
    holes of our eyes

like in a photonic crystal
in an eye-sight

this eye-light
as light as light
that doesn't know what it's doing
that cannot fail
always right
common property
"sweet
blind
selfish
light"

Yoko Seyama, artist

SAIYAH

PRISM COLOUR PLAY

Scenographic light, sculpture and choreography with ensemble
electronics music by Benjamin Staern.
Premiere at Norrlands Operan Umeå Sweden, 7 May 2014.

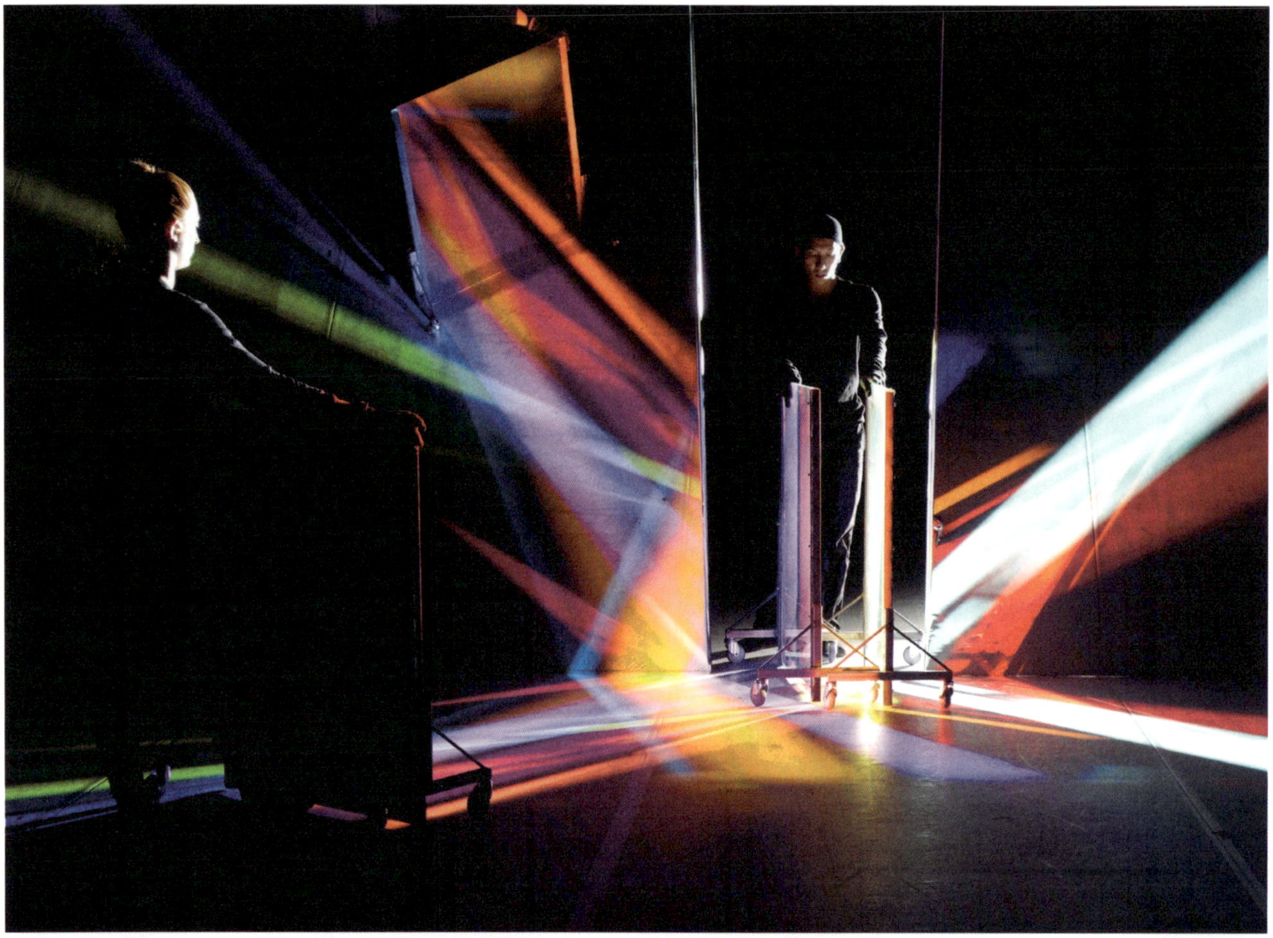

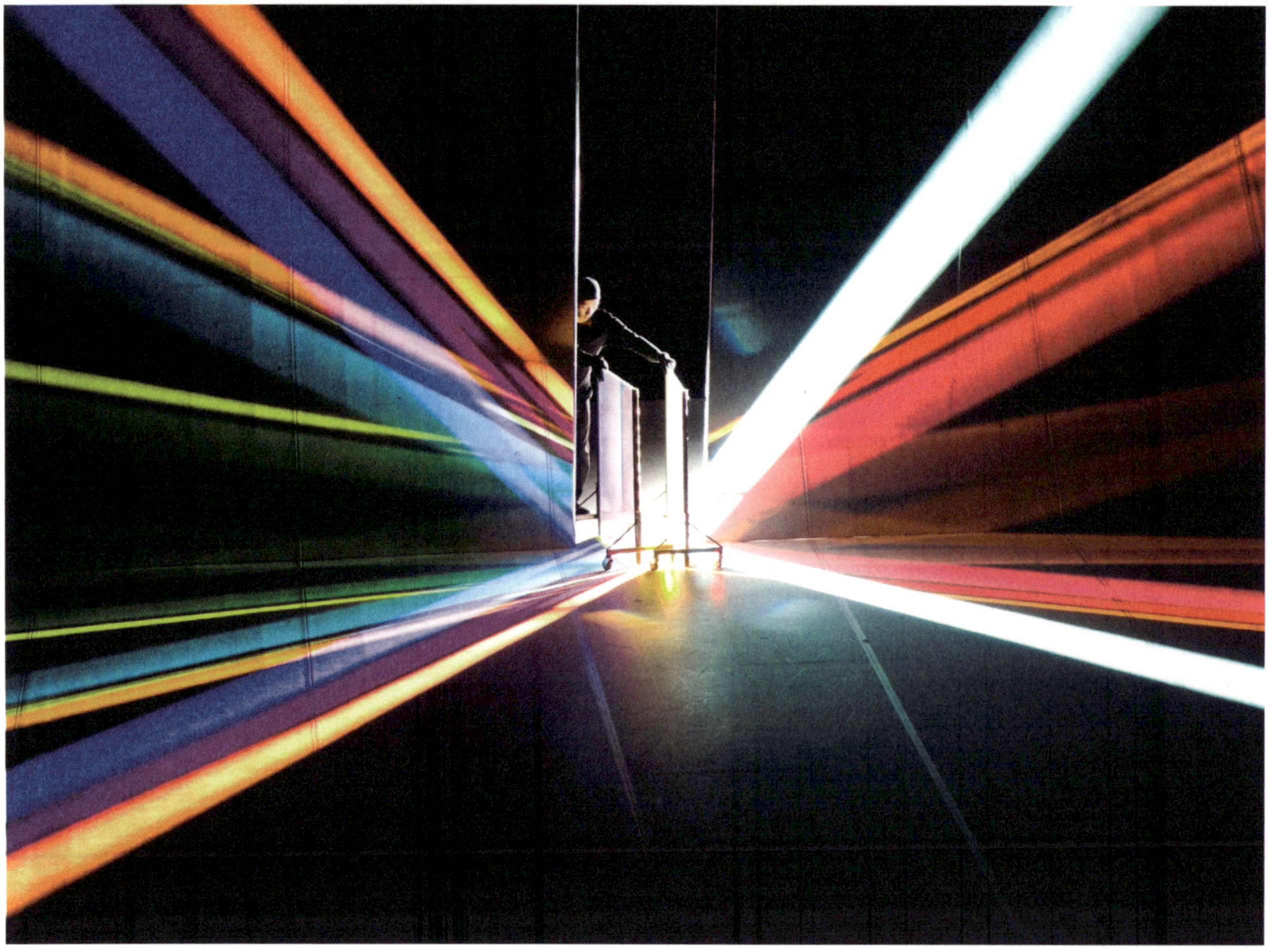

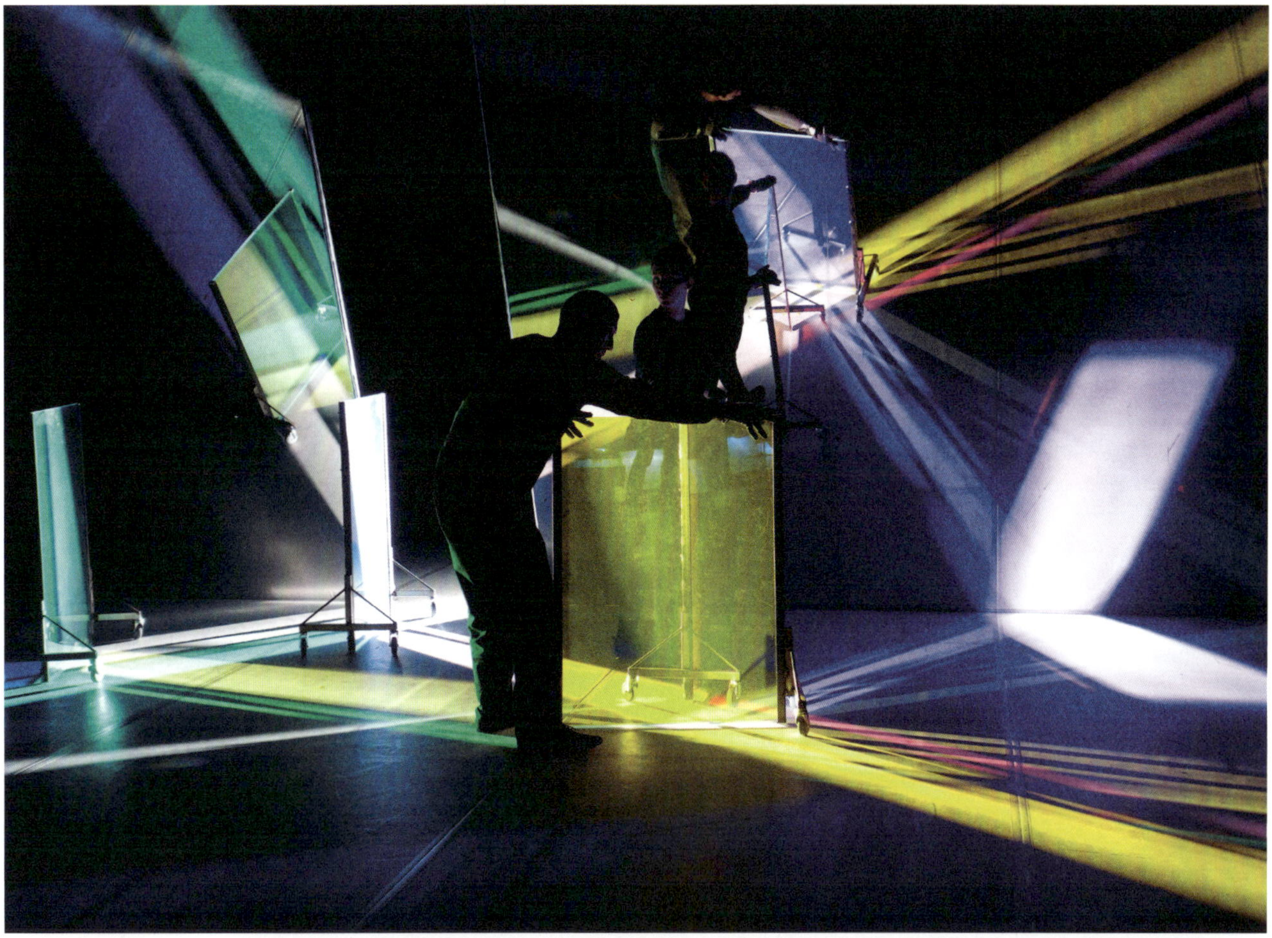

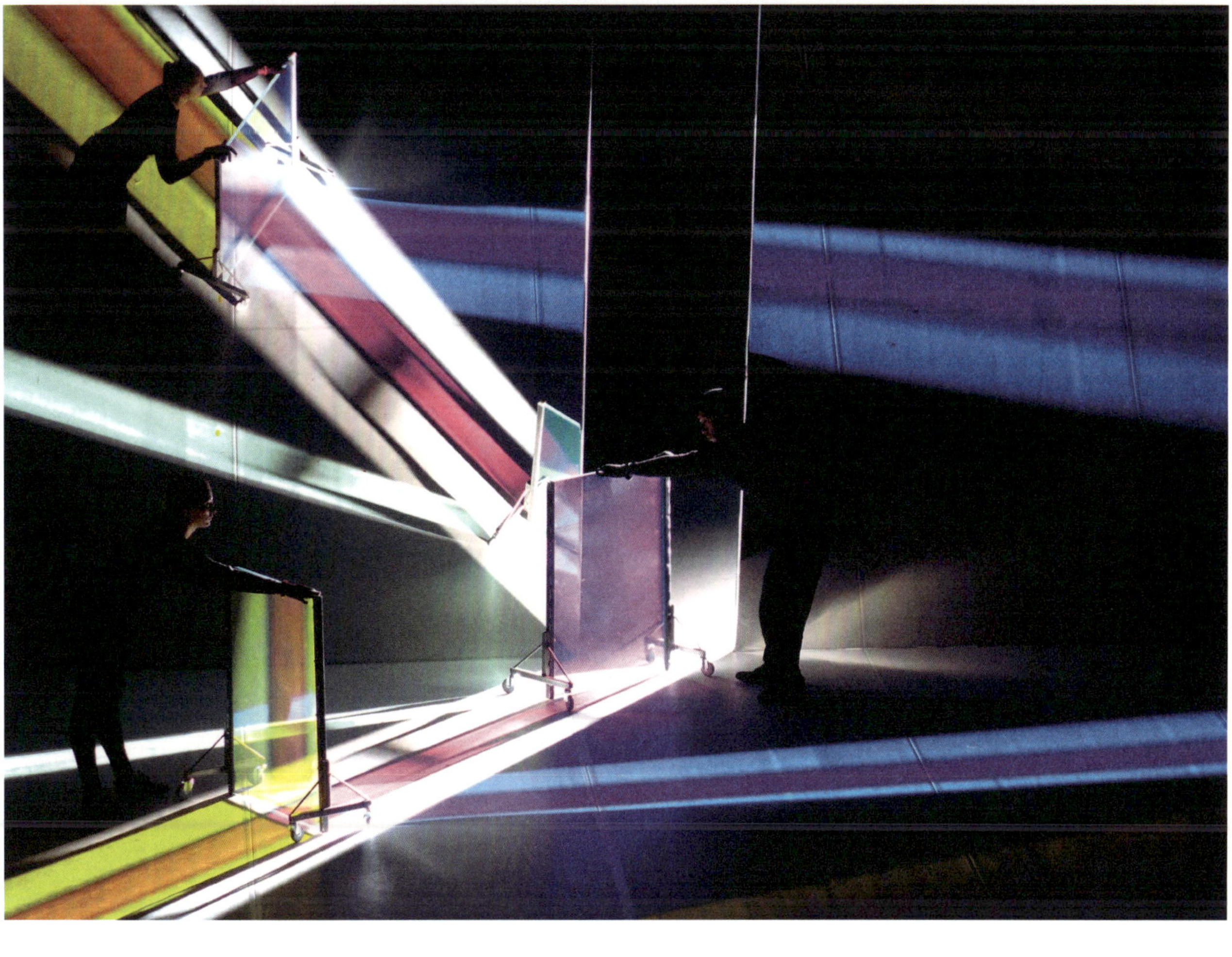

Anna Wirz-Justice, chronobiologist

enLIGHTen

CIRCADIAN RHYTHM   PHOTORECEPTORS   ZEITGEBER   MOOD   HUMAN

Light. So the world began. Life crystallized out of molecules that learned to catch the energy of the sun, transform it into building blocks of cells, organs, structures, leaves and flowers. Cyanobacteria created the first simple chemical feedback loop that provided the cell with internal time, predictive knowledge of day and night before it occurred, an advantage for survival. Biological clocks allow the organism to be prepared for the right behaviour at the right time, to avoid predators, optimize the hunt for food, and reproduce in the most favourable part of the daily or yearly cycle. Thus, it is not surprising that early in evolution, across phyla from unicells to plants to insects to vertebrates to mammals, 24-hour clocks developed at the molecular level. These internal clocks manifest an individually precise but not exactly 24-hour rhythm ("circadian" = about a day) in nearly every measured behaviour or function, and drift out of phase with real time if left in constant conditions without signals from the external world. They need to be synchronized to the day-night cycle, and the most important synchronizing agent or "zeitgeber" (time-giver) is light.

Naturalistic daylight follows a predictable pattern of light intensity day by day throughout the year specific to the geographic location.

The organism therefore requires adequate and specialized photoreceptive mechanisms to catch the light and be able to transduce information as to time of day and time of year (day length) to the central circadian pacemaker in the brain. The eye, of course, is the classic organ for perceiveing light. But we are not talking here about classic vision using rod and cone photoreceptors to differentiate colour, movement, shapes, edges. Zeitgeber function is nonvisual, and mediated by a recently discovered class of photoreceptors containing a photopigment, melanopsin, specifically sensitive to blue wavelengths. The discovery of this nonvisual role for light has revolutionized neurobiology: light affects a cascade of functions that we never dreamed of being light sensitive. Mood, cognition, performance, sleep – to name but a few important aspects of our daily lives – depend on and react to light exposure – when, how much, how long, and what spectral composition. In addition, light has immediate and direct effects on these very same functions – alertness- and mood-enhancing and activating – it's a "wake-up drug".

The above research, and consequent treatment implications, began some 30 years ago when it was found that human rhythms could be modified by light. Previously, social signals were considered the most important zeitgeber. The idea of using light therapeutically developed remarkably rapidly following basic research on the mechanisms underlying seasonal hibernation in hamsters, whereby increasing day length – adding light – triggers the switch into summer behaviour. Light's capacity to suppress nocturnal secretion of the pineal hormone melatonin is the key. And melatonin provides an internal hormonal representation of the external 24-hour day: is it nighttime? If yes, then melatonin levels are high, so please begin with nighttime behaviour – wake up hamster (fall asleep human)!

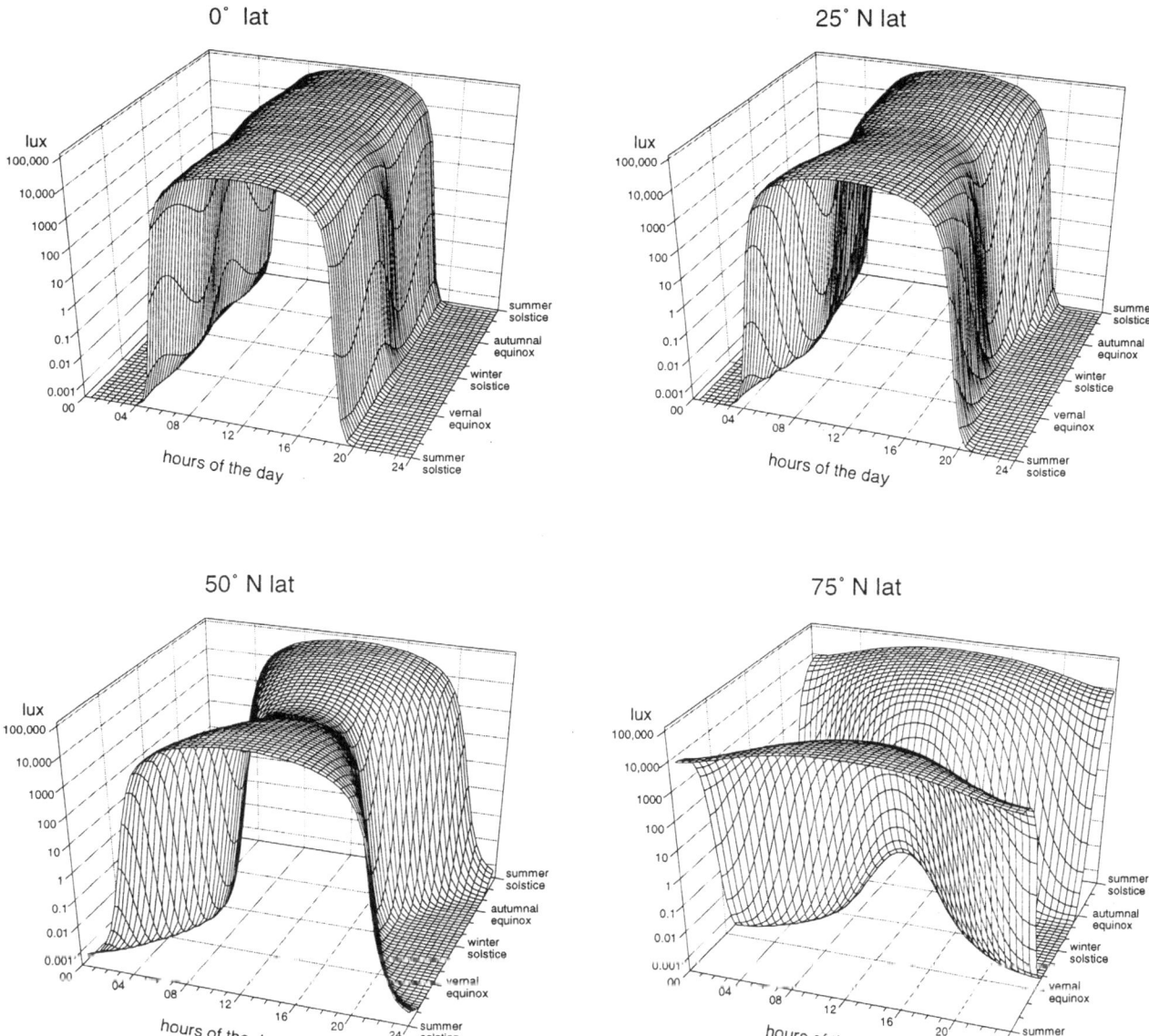

0° lat

lux
100,000
10,000
1000
100
10
1
0.1
0.01
0.001
00

04 08 12 16 20 24

hours of the day

summer solstice
autumnal equinox
winter solstice
vernal equinox
summer solstice

25° N lat

lux
100,000
10,000
1000
100
10
1
0.1
0.01
0.001
00

04 08 12 16 20 24

hours of the day

summer solstice
autumnal equinox
winter solstice
vernal equinox
summer solstice

50° N lat

lux
100,000
10,000
1000
100
10
1
0.1
0.01
0.001
00

04 08 12 16 20 24

hours of the day

summer solstice
autumnal equinox
winter solstice
vernal equinox
summer solstice

75° N lat

lux
100,000
10,000
1000
100
10
1
0.1
0.01
0.001
00

04 08 12 16 20 24

hours of the day

summer solstice
autumnal equinox
winter solstice
vernal equinox
summer solstice

01

This signal also provides a seasonal cue: long nights in winter, long duration of the melatonin signal (please hibernate); short nights in summer, short duration of the melatonin signal (please reproduce). Humans retain these neurobiological mechanisms, even if no longer overtly – they pay little attention to dawn, dusk or the seasons. So suddenly we had a novel tool, light, to manipulate the very basics of human clock function; and a brilliant jump to the clinic revealed the "pharmacological" power of light as a treatment for winter depression.

These studies began a new era: well-planned placebo-controlled clinical trials have provided a growing evidence base for the efficacy of light in many diagnoses beyond winter depression. Light improves non-seasonal depression (eg, during pregnancy, in ageing, bipolar illness), and, combined with classic antidepressants, accelerates and potentiates clinical response. Long-term treatment with light slows down cognitive decline in Alzheimer's dementia just as well as the established medication. Sleep-wake cycle disturbances in many other psychiatric (schizophrenia, borderline personality), neurologic (Parkinson's disease, post-stroke), and internal medical (cirrhosis, renal transplant) disorders can be improved with carefully

01   Contour maps of daylight patterns at four different latitudes from the equator to the far north. The 24-hour light-dark cycle (logarithmic light intensity from starlight (0.001 lux) to midday sun (100 000 lux) is shown for each day of the year.

timed light therapy. Light is an important tool in sleep medicine to resynchronize abnormal rhythms such that nocturnal sleep is re-established at the appropriate time. Widespread acceptance of such an ubiquitous and obvious factor in our environment as a serious medical treatment will, however, take time: light can't be patented, or squashed into a pill, so the limits to profit may hinder innovation

Human Retina and its Photoreceptors

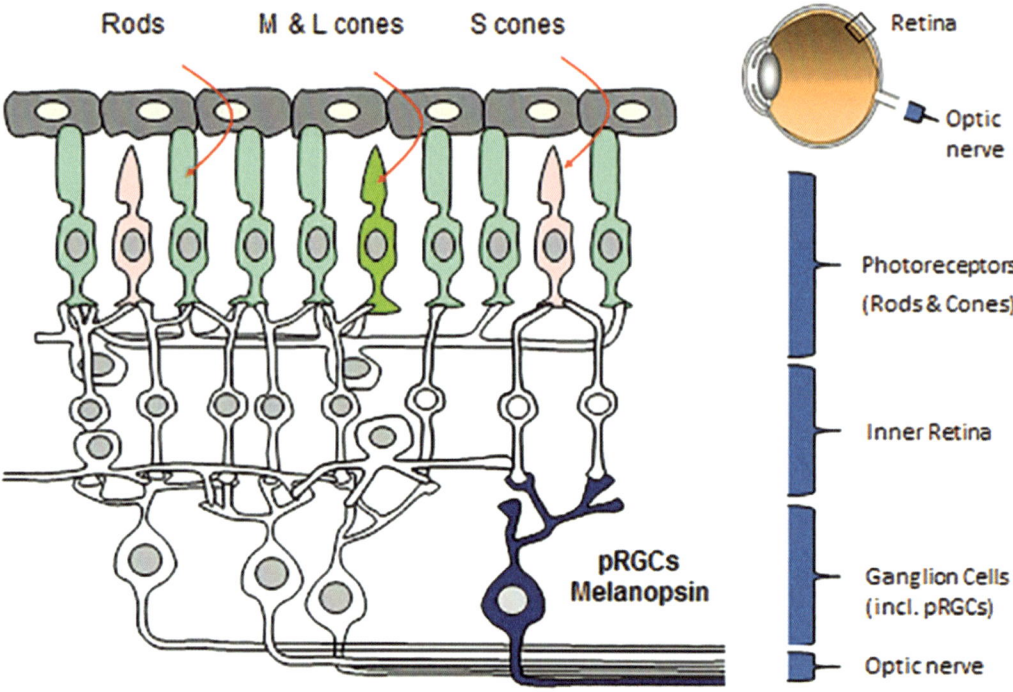

02

and application as a first-line therapy. Additionally, the image of an alternative "soft" therapy keeps doctors sceptical even though paradoxically, patients love light for that very reason – because it appears "natural" and is not a drug. Interestingly, light acts on the same brain areas as antidepressant treatment, suggesting common mechanisms of action.

Let us make an historical sidestep. The invention of artificial light broke down the barrier of night: humans could choose their wake and sleep schedules according to their fancy. But what did it do to biology? In modern society, humans no longer follow the changes of light along the seasons and live, in effect, on a constant day length. Yet we spend the greater part of the day indoors, where artificial light intensity is far below that required to reset and stabilize the clock – even though quite adequate for our visual system. We are only slowly recognizing that we are not as free as we behave – there may be health consequences of irregular, inadequate and poorly timed light cycles. And we are not even talking of social jet lag – the incongruence between internal time and social demands – that many individuals suffer, or early-bird and late-owl chronotypes, or shift workers. No longer being synchronized to the naturalistic dawn-dusk signal may contribute to vulnerability to mood and sleep disorders. Disrupted and poorly timed sleep has been linked to obesity and the development of diabetes. Even gut microbes have circadian rhythms that are synchronized by the biological clock of the host in which they live. Altered circadian rhythms in the host change the rhythms and composition of the microbial community, which may

02   The human retina and its photoreceptors. The rods and cones are image-forming cells. Rods are specialized for detecting motion and nighttime vision. The cones are responsible for colour and high-resolution viewing, some respond to long and medium wavelengths (M & L cones) or short-wavelength light (S cones). The photoreceptive ganglion cells (pRGC) are non-image-forming and respond to light intensity, due to the photopigment melanopsin. It is through these cells that our biological clock keeps time with the external world, by interpreting changes in light over periods of time.

lead to serious metabolic problems. So the molecular clockwork in all cells and organs, from liver to brain, needs adequate synchronization to maintain optimum function of the entire organism. Even our stomach bugs.

I think we are at the beginning of an "enlightened" revolution. Not only the widespread clinical implications of rhythm disruption, but also the converse realization that good light can support health. Architects have always considered light as a tool in the design of a building – even when they did not have the present-day knowledge of its deeper effects on the human body: "*L'architecture est le jeu savant, correct et magnifique des volumes assemblé sous la lumière.*"[1] Lighting manufacturers have also recognized the importance of the non-visual component of lighting, even though we still don't know enough, as scientists, to provide a simple recipe to the engineers for building healthy lights.

But we are working on it, and one day soon we will have a circadian house and live happily ever after.

1   "Architecture is the masterly, correct and magnificent play of volumes brought together in light." Le Corbusier: *Vers une architecture*, 1923

Bradley Pearce, neuroscientist

THE WALL IS NEVER YELLOW

COLOUR CONSTANCY   PERCEPTION   WAVELENGTH   BRAIN   SUBJECTIVE

The tremendous melee of light that is reflected from and between objects, from light sources such as the sun, is a colourless fog. Light has no colour, hence you can see through this fog. Instead you see surfaces at relative depths, of different colours, which are all created by the brain in response to the signals passed to it through the eyes.

Imagine a wall to your left, one that you cannot see as you are gazing straight ahead. Light reflecting off that wall is passing straight past you, yet you are unable to see it, as the light does not enter your eyes. As you turn to face it, the photosensitive cells in your eyes catch some photons and then you see an image, which is an interpretation by the brain of the information conveyed by light bouncing off the wall.

Why do I say interpretation? Well, we see not what is really there, but what our ancestors needed to see to survive. Imagine that same wall with a shadow cast upon it, you can effortlessly tell what is shadow and what is surface, what the surface would look like without the shadow, the texture and the colour. We can tell the colour of the light and the colour of surfaces, even though this should be impossible to truly determine, as the light entering the eye is the multiplication of the incident light and the surface reflectance. Indeed, we so effortlessly do this that we attribute properties to those surfaces, like saying that the wall is "yellow" when painted with a material that reflects mostly medium- and long-wavelength light (most people call this material yellow paint).

The wall, or any surface for that matter, will reflect more short-wavelength light under the pale-blue light of natural skylight or more medium-wavelength light under the yellow of direct sunlight, yet we still see the surface as having roughly the same colour. Indeed, object colours seem to stay roughly stable as the illumination upon them changes, a phenomenon called colour constancy. The mechanisms the brain uses to achieve colour constancy work so well that we give objects colour names – like red fruit, and green leaves; a useful ability for our distant savannah ancestors despite colour being a complete construction of our brains rather than a property of the world.

When our brains are not colour constant, we tend to perceive the colour as if it is lit in isolation. This can be quite shocking, as we are so used to object colours staying stable. This is one of the reasons camera images can be disappointing, as cameras do not have our ability to automatically compensate for changes in chromatic light on the scene; regularly rendering images too blue or yellow. This can be even more shocking when the lighting on surfaces is ambiguous, and the brain doesn't quite know what colour to make of the surface. This was the case with the infamous dress photograph that circled social media in early 2015; most people reported that it looked blue and black, but others reported it as white and gold. Some people thought this was the result of their screen setting but this is not the case: suddenly there was mass-awareness that colour perception *really* is subjective.

You might be thinking how perception can vary between people so radically? A surface reflecting lots of blue light *could* appear white if you believe the light on the scene to be blue, or appear blue if you assume the light to be white.[1] The really fascinating thing is that different people appear to be using different information from the image to construct their perception of the dress, and no one knows what puts your brain in the blue/black or white/gold camp. Such a trivial image gives us a rare glimpse that our reality is entirely constructed by that electro-chemical mass in our heads; even so, we still find ourselves asking, what colour is the dress really? Our desperate attempt to assert reality as the way we perceive it leads us to ask what colour the dress would appear under perfectly white light. I will leave it to you to decide whether there is an answer, and if so, if it is satisfying at all.

At the invitation of Astrid and Hester from PARS, Professor Anya Hurlbert and I constructed an experiment as part of their Some Like Dark show at the Wellcome Collection in London in 2015. We used a physical dress the same as that in the infamous picture, and LED light sources from our Hi-LED, with which we could specify the exact spectral content of the light that was emitted from them. We carefully controlled the light to create conditions whereby the dress surface colours were ambiguous. Audiences were asked to fill in a questionnaire and name the colour of the dress under these lights: around three-quarters of participants said that the dress was blue and black and the remainder said that the dress was white and gold (among other combinations like brown and blue). Yet, when we lit the dress using perfectly white light, almost everybody agreed that the dress was blue and black.

Under regular lighting conditions we can mostly agree on colour names. Interestingly, the subtle differences in our colour perception, born from individual differences, the age of our eyes and our environmental history, are revealed under carefully constructed experimental conditions.[2] Under these conditions, older observers rate bluer light more white than younger observers due to more short-wavelength light being filtered by the optical equipment. Most interestingly, these differences are rarely observed in ecological conditions; notably, there were no differences of age or gender observed for the colour names given to the dress in the experiment at the Wellcome Collection. This means that there must be some compensatory mechanisms to allow our view of the environment to remain constant, despite the degeneration of our eyes.

Our ability to generally agree on reality allows us to communicate, and so uncertainty about the true properties of objects can cause the variety of reactions produced by the dress photo. Perhaps this is why reactions to abstract representations of our shared reality can be so diverse – after all, the coloured forms created by the artist out of spots of pigment on canvas are no more real than on the objects that inspire them.

1   D.H. Brainard, A.C. Hurlbert (2015), "Colour Vision: Understanding #TheDress", *Curr Biol*, 25:R551–R554
2   S. Werger (2013), "Colour Constancy Across the Life Span: Evidence for Compensation Mechanisms"

THE MIND'S EYE

EYE-LIGHT | SPACE | SHAPE | MANIPULATION | PERCEPTION | MYSTERY | PLAY

If there is anything in our lives that is truly in the eye of the beholder it is light; its very existence is hard to prove. It is reflected from an object into our eye and we see the object, not the light. It is no wonder that there are so many people who feel that they know so little about light. To me it is the one mystery of our lives that refuses to reveal its secrets. Yet we humans take it for granted. It is our birthright to see and we wouldn't be able to if there were no light. From the beginning, painters tried to capture its mystery in two dimensions; stage designers tried to discover its secrets in the theatre. It is profoundly truthful in what it reveals; yet it can hide as much as it shows; it lies as often as it reveals the truth. It is no wonder that many are suspicious of its illusive and illusionistic qualities. However, it is a miracle that the patterns of light and dark – black, white and colour – can be perceived by our minds as moving objects in space. It is interesting to me that now that we have begun to play games with the idea of virtual reality we begin to realize that all reality is virtual – that what we see is what we have in our mind to see in the first place.

In some sense there is as much a reflection from our mind to the object, allowing us to "see" it, as there is a reflection from the object to our mind through the eye. In that sense, all light is a form of reflection.

When I was starting out as a lighting designer, I had an extraordinary experience that has stayed with me all my life. I was in Italy with the Paul Taylor Dance Company and we had some time off, so a group of us went to Florence to see the sights. The only one of us who could speak a little Italian was a very beautiful young woman – a dancer, graceful, soft-spoken, shy. We four young women went to the Academia to see Michelangelo's David. Liz asked the guard for directions. The guard, an Italian, remember, immediately fell in love with her and became our personal guide, bidding us, once we came to the statue of David, to feel as much as we wished and as far as we could reach. That experience totally changed my way of seeing.

When I touched that wonderful calf, knee and thigh I expected to feel cold stone. Much to my shock I experienced what I could only call flesh and bone. The shape as it fit into my hand sent my brain signals that it was what it was shaped to be, somehow discovered buried in that marble. It seems to me that everything we make must be uncovered, dug out of the stone to be revealed in the light. It's not easy work to separate human form from the stone, or from the darkness, but isn't it somehow the work of the artist in his or her pact with an audience to reveal what is hidden there?

As a lighting designer I am constantly grappling with finding the way – the form, the style – that will parallel the work of the others involved in the production to lead the audience to those places and states where insight can come.

For me light is the music of the eye. Light has the same mystical power as music to transport an audience effortlessly from one place, one moment, one idea to another without having to make the individual steps along the way. It has range, variety, dynamic; it has colour; it has volume; it has rhythm. I believe that, through the formal use of light, we can shape the way that we see the world and therefore the way that we think about that world. Light can make for us a reality that is not literal but whose substance we do not question. State the theme and the variations are endless – the stricter the rules the farther they allow one to go. I feel lucky to have spent my life struggling to understand the mystery of this extraordinary stuff of no substance that "dreams are made on".

Adam Fuss, photographer

UNTITLED

FOG SOFT ENDLESS

Unique gelatin silver print photogram, 2005
61 × 50.8 centimetres

Robin Dunbar, anthropologist, evolutionary psychologist

A LIGHT MOTIF:
THE FORGOTTEN STORY OF
HUMAN EVOLUTION

CAMPFIRE   STORY   EVOLUTION   PLAY   ROOTS AND TUBERS   EYE SOCKET SIZE

Like all monkeys and apes, we humans are a tropical species. We have deep origins in the tropics and their rich, luxuriant forests. The advantage of the tropics is that the climate is relatively consistent – no summers or winters, just something closer to spring and autumn. And more particularly, the pattern of night and day remains resolutely constant. The day is a steady 12 hours, the night exactly the same. There is no variation from one season to another across the year. Those of us who live in the northern and southern extremes of the planet know only too well how day length changes from summer to winter. Indeed, this becomes so extreme above the Arctic and Antarctic circles that the entire 24-hour day is daylight in summer, and the entire 24 hours are night in winter. For high-latitude species, winter is a trial not just because it is cold but because the time available in which to find food is very limited – unless they can do this as easily at night as in the day.

Like all monkeys and apes, we are daylight species: our eyes are poorly adapted to the dark because we have replaced many of the light-sensitive cells in our eyes that respond to black-and-white with cells that respond to colour. We are all but blind in the dark without the help of a full moon or a street lamp. In the tropics, where our ancestors lived, this wasn't too much of a problem: a 12-hour day was pretty much adequate for doing all that was necessary to keep body and soul together and doing all the social things necessary to keep our social groups coherent. Usually, there was plenty of time left over during daylight for some relaxation, a quiet doze during the heat of the day when the fierce tropical sun makes it rather uncomfortable to be busy.

This was all very fine and idyllic. And so long as our ancestors remained in ape mode (which they did from their first appearance around six or seven million years ago until about two million years ago), this wasn't a big deal. The larger brains and bodies that are typical of the apes certainly put some pressure on our time budgets, but we could just about squeeze everything we needed to do into a 12-hour day. But, some two million years ago, our lineage evolved as a new genus characterized by ever-increasing brains and larger bodies. Things began to get a great deal tougher.

Indeed, things were so close to the limit of what we could manage that further evolutionary change, and in particular the evolution of our very large modern brains, would have been impossible. We would have been locked down at the level of what we might think of as a smart ape-man, but nowhere near as smart as we are now. There would be no fine architecture or rousing music, no Mozart or Wagner, no Shakespeare or Goethe, no comedy or jokes, in fact no language in which to do all these very human things. Something dramatic had to happen to allow us to break through what was a very serious glass ceiling. That something was the invention of a new source of light.

HOW FIRE MADE BEING HUMAN POSSIBLE

Most people who think about fire in the context of human evolution think of fire almost exclusively in the context of cooking. Of course, cooking, once we had invented it, did play an important part for us. You might well say that it made food taste better, and it certainly made some foods more digestible. Though by no mean indigestible, meat, for example, is pretty tough stuff when raw; but roast it over an open fire, and its digestibility increases by about half again. The same is true of most roots and tubers. Think of your average potato. Eat it raw and it is pretty grim, if not indigestible; but boil it up or roast it and it becomes an altogether different thing. The advantages of figuring out how to cook your food are very quickly obvious. What's more, because cooking increases the digestibility of many foods, it reduces the amount of food you have eat to satisfy your nutritional needs, and hence radically reduces the amount of time you need to spend finding it. Fires have another advantage, of course, especially once you get into cold climates. Having a fire at night has obvious benefits, especially in a cave: it greatly increases the temperature. On the cold plateaux of Africa (where nighttime temperatures can drop as low as 50°C) and the bracing winters of Europe and northern Asia, sleeping round a fire can mean the difference between survival and extinction.

However, fires have another important benefit that is almost always overlooked. They provide light. By sitting around the fire at night we can, in effect, extend the working day. Monkeys and apes take to their beds as soon as it gets dark; they have no choice, because their poor night vision means they cannot carry on feeding during the hours of darkness, and they certainly cannot afford to move about, looking for food, for fear of being caught unawares by a predator. Instead, they retreat into the trees, find a nice nook in the canopy and hunker down for the next 12 hours until dawn, mostly dozing fitfully. For several million years, our ancestors were forced to do much the same, but probably mostly in caves – we lost the ability to clamber around in trees when we started to walk upright, around seven million years ago. Nighttime was dead time.

But if you can control fire, and keep it burning all day and into the night, then not only does it keep the place warm, but it provides light during the evening hours. That, in effect, extends the hours of daylight into the night. Now, there are clearly some things you still cannot do. You cannot, for example, wander off, out of the circle of light, to start foraging, so you are a bit stuck within a very small spatial area. And when all you have is logs and branches to light, you cannot exactly carry a lamp with you to show you the way if you do want to wander off. However, if you can switch at least some of your more sedentary activities into the evening hours, that would free a lot of the daytime for those activities like foraging and travel that have to be done away from the cave.

The two obvious activities that you can do round the campfire in the evening are, of course, cooking and eating, and making stuff. Once we had evolved efficient hunting (some time around 500 000 years ago, judging by the evidence), there was a need not only to cook lots of otherwise semi-digestible meat, but also to make and repair the weapons we needed to do it with – spears, spear-throwers, bows and arrows, protective clothing, etc. Making and repairing tools and kit probably did not require much imagination, given that you were forced to sit around the hearth in the evening doing nothing more than twiddling your thumbs. However, cooking and eating in the evening required us to radically change our style of living. Like most animals, monkeys and apes eat as they go along. For most of our evolutionary history, we did the same. The switch to eating by candlelight, however, was a major innovation, and it is still with us.

There is something magical about eating in the semi-dark in the flickering light of a fire or candles. We are so in its thrall that we deliberately engineer our social eating this way. How often do we ask people round to breakfast, or even lunch? If we really want to do the social thing, it's always dinner. Eating in the half-dark adds a significant mystique, makes the whole experience so much more intimate. And this is not because we are busy grubbing in the fields during the day and don't have the time to invite friends to lunch. We still do it at the weekends when we are free of all that forced labour and could lunch if we really wanted to. We might do a business lunch, but that doesn't have quite the social caché of dinner in the evening. Even at the weekend, we much prefer to ask people round to dinner. I suggest this is some kind of hangover from the dim and distant past when we took to having our major feeding time in the evening – at the end of a long day of hunting and gathering.

Why is this important? The answer is that feeding time can become social time. For monkeys and apes, feeding time and social time are quite separate things. The most extreme case of this is the gelada baboon, which I spent many years studying in the wild in Ethiopia during the early part of my career. Gelada sleep on massive cliff faces, and come up to the cliff tops once the sun has warmed the air. There, they spend as much as two hours engaged in social grooming. Only once they have done this do they start feeding, and they then spend the rest of the day feeding more or less without interruption. We do the reverse: we spend the day foraging (or, alternatively, working to earn the money to buy groceries ...) and then devote the evening to socializing. That allows us to maximize the time we can devote to foraging (or working) during the day, unlike the gelada, who lose several hours from their foraging day by having to devote time to social activity.

What we have managed to achieve here is remarkable in two respects. One is that we have managed to postpone feeding for much of the day. This doesn't mean we don't feed at all during the day, but it does generally mean that we reserve the big meal for the evening and something more like a snack for midday. Putting off our big feed rather than eating continuously through the day, as monkeys and apes do, is,

in fact, no mean feat. It depends on our being able to postpone immediate reward in the interests of a bigger reward later. That's something that monkeys, apes and children find hard to do: present them with the choice between a cake now and a much bigger cake later, and they will grab the small cake now. Being able to postpone reward in the interest of getting a bigger reward later in the way we do requires an extra bit of the brain in the frontal lobe just above the eyes. This bit is not unique to humans (all monkeys and apes have it), but it is unusually large in adult humans, and it allows us to postpone immediate reward for much longer than any of the monkey and apes.

The other respect in which postponing feeding until the evening is remarkable is that eating becomes a social rather than an individual event. Monkeys and apes nibble steadily through the day, but it really doesn't matter whether there is someone else around or not. But for us, an evening meal is an intensely social occasion. We deliberately wait to eat until everyone else is ready. That allows us to use eating time as social time, as time to bond and build friendships. Eating socially becomes more like a conversation or a dance, and in doing so it helps us create and maintain friendships that are crucial to our very survival.

The impact that shifting social time to the evening had on the time budgets of our ancestors is dramatic. Figure [01] shows the estimated time budgets of the main species of ancestral hominins (the members of our lineage). Let me briefly explain how we obtained these estimates. We have developed models of the time budgets of the two African great apes, the gorillas and chimpanzees. These time budgets were determined from actual data on time spent feeding, moving, resting and socializing in these and other primate species. Time budgets are dependent not just on a species' physiology but also on the climate of the particular habitats where its populations live. So what the model provides is a set of equations that relate each of the time budget components (feeding, moving, resting and social time) to local temperature and rainfall variables. With this model as a baseline, we then extrapolated the equations to derive time budgets for the australopithecines, the earliest members of our lineage. The time budget we arrived at for the various australopithecine species is very tightly delimited by the time available in a 12-hour day: australopithecines had no free time to play with at all. If we take this as representing 100% of the available time, we can then extrapolate to the time budgets of the later species of ancestral hominins, all of whom belong to the

01 Median (and 50% and 95% ranges) of percentage of a standard 12-hour day that is required to meet the time demands for feeding, travel, resting and socializing in each of the major species in the lineage leading to modern humans. The dotted line indicates 100% of the day. The solid line indicates the equivalent when an additional four hours is added to the waking day by using fire to create artificial light in the evening.

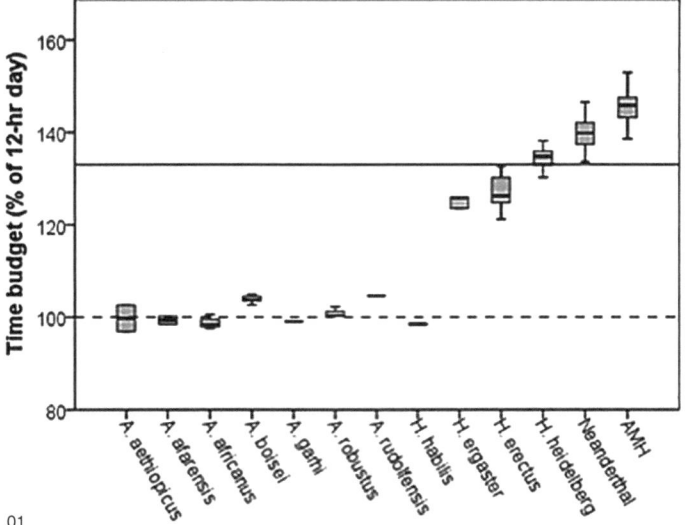

for socializing to maintain the cohesion of their communities. The result is what we see in [01].

The key message is that all the later species after the australopithecines would have had an increasingly difficult time balancing their time budgets within the limits of a 12-hour day. If we were to service our relationships in the same way as other monkeys and apes, namely by social grooming, then our time budgets would be about 50% over the limit we would actually have time for in a 12-hour day. We could, of course, have smaller social groups; that would have meant we could have got rid of quite a lot of our excessively large brain, and saved a lot of extra feeding time as well as requiring a lot less time for socializing. That would certainly have reduced our time budget requirements, but it wouldn't have eliminated the overrun completely, because, in effect, it would only have taken us back to the position of the early members of the genus *Homo*, such as *Homo ergaster*. More importantly, in doing so we would have had to sacrifice all the major cultural achievements that depend on our large brains – language, religion, literature, science – not to mention our knowledge of coping with difficult environments. We would have been back to square one.

Looking again at the pattern in [01], it seems that we might just have been able to deal with the relatively modest overrun of about 20% faced by the early species, *Homo ergaster* and *Homo erectus*. They seem to have coped with this by a combination of an improved diet and adopting laughter as a social bonding mechanism. Laughter is still important for us in this context, and has all the hallmarks of being very ancient. The real problem would have been coping with the massive further increase that comes with the later species – archaic humans (*Homo heidelbergensis*), the Neanderthals (*Homo neanderthalensis*) and our own species, anatomically modern humans (*Homo sapiens*). It is here that fire as a source of light comes into play.

The archaeological evidence demonstrates that, although fire was occasionally used for cooking from about a million years ago, hearths only became a regular feature of human living sites from about 400 000 years ago, soon after the first appearance of archaic humans (*Homo heidelbergensis*). This date marks a dramatic change in the presence of hearths at fossil sites: before this date, they are found at less than 10% of all sites, but after this date they are found at around half of all sites.

Adding another four hours on to the active day by extending the waking period into the evening would have dramatically reduced the

new genus *Homo*, our immediate family, which appeared for the first time around two million years ago. We can do this by scaling the feeding and socializing requirements by the ratio of each species' brain and body size to the average size of australopithecine brains and bodies. Feeding time is determined by how much time is needed to fuel your somatic tissue (your body mass as a whole) plus the additional extra cost of fuelling a larger brain (because brain tissue is about ten times more expensive in terms of energy consumption than muscle tissue). In primates in general, social relationships, and hence group or community bonding, are mediated by social grooming, which increases more or less linearly with social-group size across the monkeys and apes. Social time is thus basically grooming time, and nothing else. The size of the brain also tells us how big the social communities of these species was, because brain size is related to group size across the primates (including humans). So we can calculate exactly how much time they needed

162

time costs our ancestors faced. A 50% overrun on the time budget such as that calculated for anatomically modern humans [01] is equivalent to finding an extra six hours of daylight. Adding four hours round the fire would change the baseline of 100% of the waking day to the position of the solid line: it would increase the waking day by a third. That would have restored archaic humans and our modern human ancestors to the position of the early members of our genus. It would have required us to find savings of only about 15%, and this had already been engineered by their predecessors through a change of diet and the use of laughter for bonding. We had broken through the glass ceiling.

Spending time sitting round the hearths of an evening turns out to have another advantage. Rather than sit in stony silence, it provides the perfect opportunity for conversation. And that conversation has to be a vocal one. Many have tried to claim that human language evolved first as a gestural language, usually because those who advocate this position assume that we use language mainly for hunting and a gestural language based on hand signals would interfere less with the hunting. In fact, most hunting is done in silence: hunters know exactly what they need to do and need little communication among themselves to achieve their objective, just as most pack hunters do. But in the dark, around the campfire, gestures have a very limited use. You can certainly wave your hands at the person next to you, but the person across the fire can barely see what you are signalling; and the people at the next fire cannot make out what you are saying at all, even if they realize that you are trying to communicate. But shouting out what you want to say allows everyone to hear. So maybe language as we know it evolved around the campfire.

But once you have language of this kind, it suddenly offers a novel possibility. You can tell stories and entertain with jokes in a way that simply is not possible with a fully gestural language. Storytelling is important because it is shared stories of this kind that provide the basis for friendships and for bonding communities. Sharing the same worldview allows us to realize why we belong to the same community, why we share so much in common, where we came from and so on. It also provides the basis for the evolution of fictional storytelling (and hence, eventually, literature as we know and love it) and, of course, the evolution of religion. Indeed, a recent study of conversations by Bushmen in Southern Africa revealed that most evening conversations involved storytelling – tales of the old days, accounts of trips to distant places,

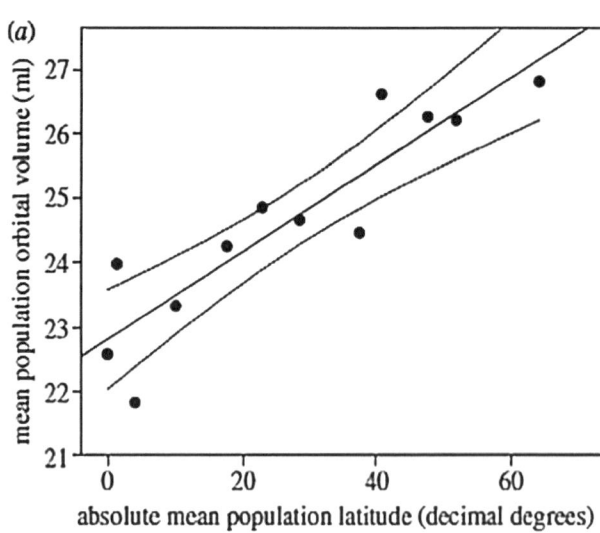

02

recollections of the past, etc. Daytime conversations, in contrast, tended to be more factual and deal with matters of immediate economic importance. It seems that there is something special about the evening when it comes to storytelling. As with eating, a story in the half-dark round the campfire is more exciting and transporting than the same story would be in daylight.

In short, the use of fire to extend the day made possible the late upsurge in brain size that ultimately made modern humans, and so gave rise to that cultural explosion that made us what we are today. Without this, it is unlikely that any of the several human species alive some 200 000 years ago would have been able to invade northern Europe and Asia. The demands of the cooler climate of central Eurasia, especially as the Ice Ages kicked in from around 500 000 years ago, combined with the greater unpredictability of food supplies that this necessarily triggered, meant that colonizing high-latitude habitats was only possible with the greater behavioural flexibility provided by the large brains of these later species. The early, smaller-brained members of the genus *Homo* (*Homo ergaster*, *Homo erectus* and their Georgian cousins *Homo georgicus*) never colonized Eurasia north of latitude 41o°N, whereas big-brained Neanderthals and anatomically modern humans were able to colonize sites as far as 50o°N, approximately 1000 km further north, despite the fact that the climate they faced was considerably colder.

THE LIGHT OF THE (NORTHERN) WORLD

The occupation of significantly higher latitudes by our later ancestors brought with it an entirely different, and in many ways unforeseen, problem. High latitudes at both ends of the earth not only suffer from extreme seasonable variation in day length (mid-winter day length at 50o°N is less than eight hours, compared to twelve hours in the tropics), but they are also subject to more cloud cover and significantly lower light levels. Having to cram all one's essential activities into four hours less than are available in the tropics inevitably puts an enormous strain on a species attempting to invade high latitudes. And the problem is compounded by the fact that, even when you can see what is going on, visibility is much reduced because, thanks to the curvature of the earth, sunlight has to travel through a much thicker layer of atmosphere. Spotting prey animals to hunt or predators to avoid just becomes that much harder.

It turns out that both the Neanderthals and anatomically modern humans adapted to these circumstances by increasing the size of their visual system to allow more acute vision under the prevailing lower light levels they faced. Figure [02] plots data for modern humans based on museum specimens dated to within the last 300 years. The higher the latitude north or south of the equator, the larger the population's eye sockets were. It turns out that this increase in eye socket size parallels an increase in eyeball size and in the size of the visual areas at the back of the brain, as determined from CT (computerised tomography) scans from living humans in different parts of the world. At the same time, visual acuity under ambient daylight (ie, not in the optician's chair, but outside, under natural light) does not change across latitude. In other words, populations living at high latitudes increase the size of the eyeball (in order to have a bigger retina for incoming light to fall on), and the size of that part of the brain that handles visual processing so as to compensate for declining light levels and the impact that has on our ability to discriminate between objects out there in the environment – between a deer in the bushes that you can eat and a sabretooth cat in the bushes that might eat you instead.

Both Neanderthals and modern humans paid a price for this. The increase in brain matter devoted to visual processing was achieved without sacrificing those bits of the brain required for other purposes, and in particular those brain regions (mostly at the front) that handle social skills. In other words, rather than sacrifice their social skills and the social community size these made possible, both species preferred to increase the total size of the brain and pay the extra cost of growing and maintaining the additional tissue. Given their pressured time budgets, that is no mean feat. But one consequence is that people who live at high latitudes, both north and south of the equator, have bigger brains than those who live at the equator (something, by the way, that we have known to be the case for more than 50 years). However, that doesn't mean that people at the top and bottom of the earth are smarter than those who live at the equator. All it means is that they have better vision. And that has one perhaps less desirable consequence: when those of us who come from high latitudes go to the tropics, we are more likely to be blinded by the sun and have to grab a pair of extra-dark sunglasses.

What really made this possible, however, was the fact that control over fire both allowed us to cook our food (and so reduce the time we needed to devote to foraging for food) and at the same time extended the day so as to give us that extra precious time that we could use for chatting round the campfire.

02   Mean eye socket size for different populations of modern humans plotted against the latitude at which the population lived. The data are from museum specimens dated to within the last 300 years.

Tatsuo Miyajima, artist

SEA OF TIME FOR TOHOKU
(PROPOSED MEMORIAL)

REMEMBRANCE | TIME | LIFE AND DEATH | SOUL | INFINITY

Tokyo, 29 June 2015

Dear Reader,

This sketch is an idea I have for the repose of the souls of the victims of the Great East Japan earthquake in March 2011.

15 890 people have died as a result of this disaster and 2589 people are still missing. By using 3000 LEDs, I am thinking of exhibiting the light of life of those who passed away. In Buddhism the number 3000 symbolizes the universe in its entirety.

The LEDs count numbers from nine to one, excluding the number zero. When counting numbers, it represents life, and when the lights go off, it signifies death. The sequences are repeated ad infinitum. In Buddhism, this is called "Samsara", and means reincarnation, the indefinite cycle of life and death. In other words, death is not the end; it is a preparatory period for the next life.

The counting speed can be adjusted down or up, the same way that each life has a unique character. I call this adjustment process "time setting", and I would like the family members of the victims to do this. I want them to "set time" with their loved ones in mind, to be remembered everlastingly in the land of Tohoku.

The waterproof LEDs will glow in a huge pool. Many of the victims died from the tsunami that came after the earthquake, and most of the missing are said to be in the sea. However, the bereaved families hardly seem to bear grudges against the sea, for many years people lived together with it. That is why I thought the sea was an appropriate place of repose for the souls of the deceased.

I am looking for partners who are willing to make this happen. Those who are interested in the project, please visit my website: tatsuomiyajima.com.

Yours faithfully,

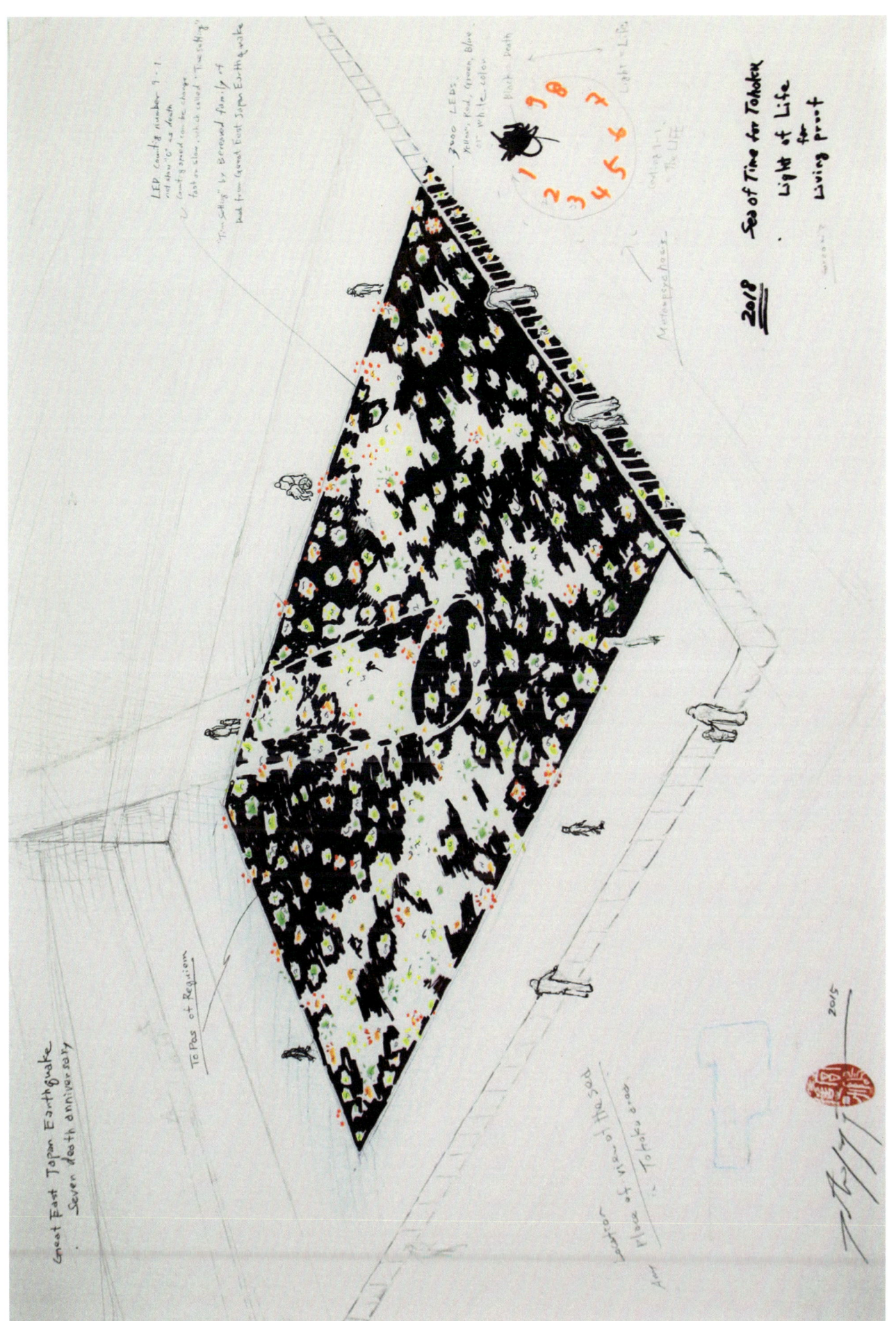

Great East Japan Earthquake
Seven death anniversary

To Pas of Requiem

Location
Place at view of the sea
— Tohoku area

2005

LED counting number 9-1
not slow "0" as death
Counting speed can be change
faster-slow which called "Twi-light"

"Twi-lighting" by Bereaved family
had from Great East Japan Earthquake

3000 LEDs,
Yellow, Red, Green, Blue
or white colour

Black Death

Light—Life

Light-Life
The LiFE

Metempsychosis

2018   Sea of Time for Tohoku
        +
        Light of Life
        for
        Living proof.

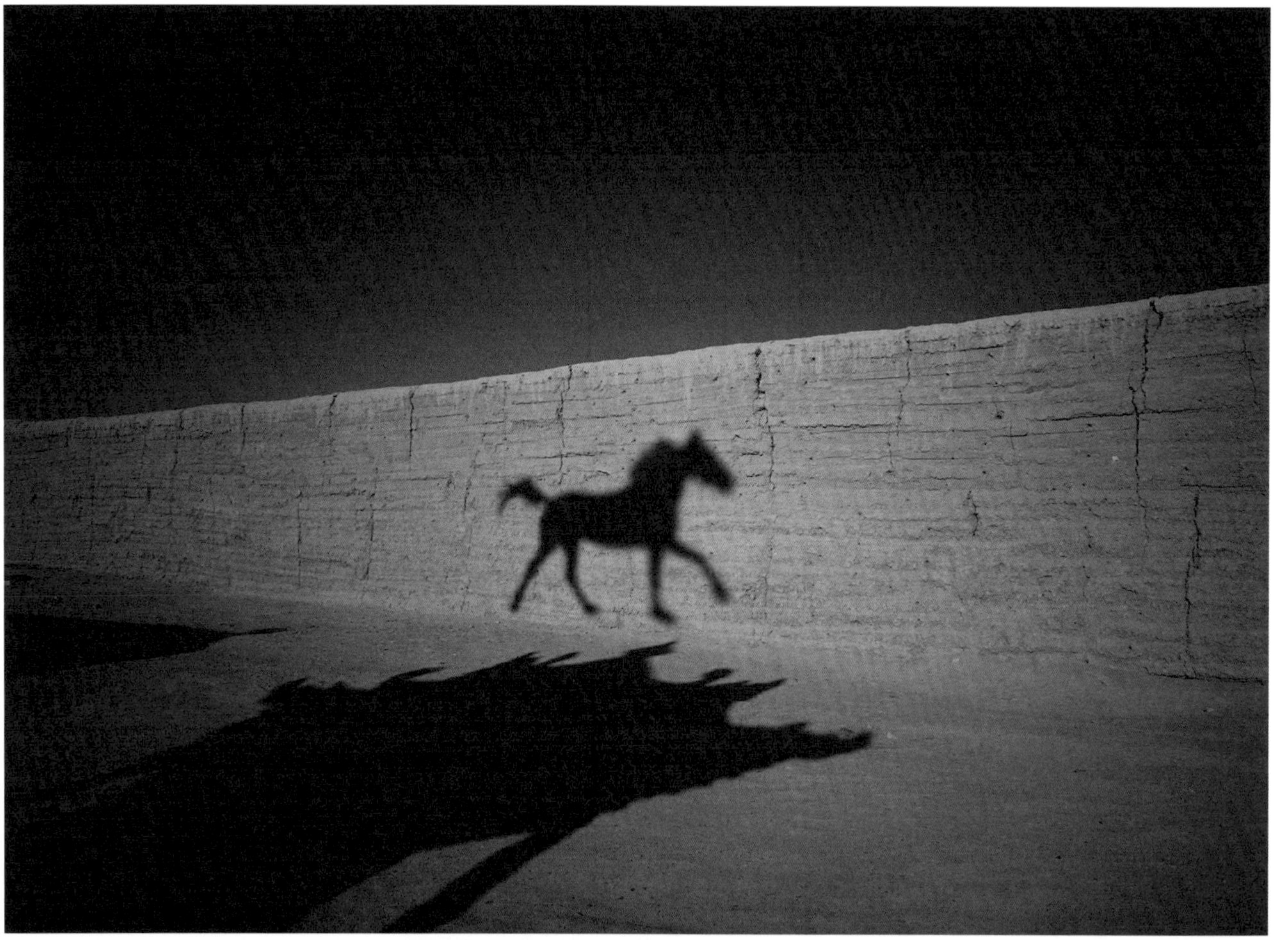

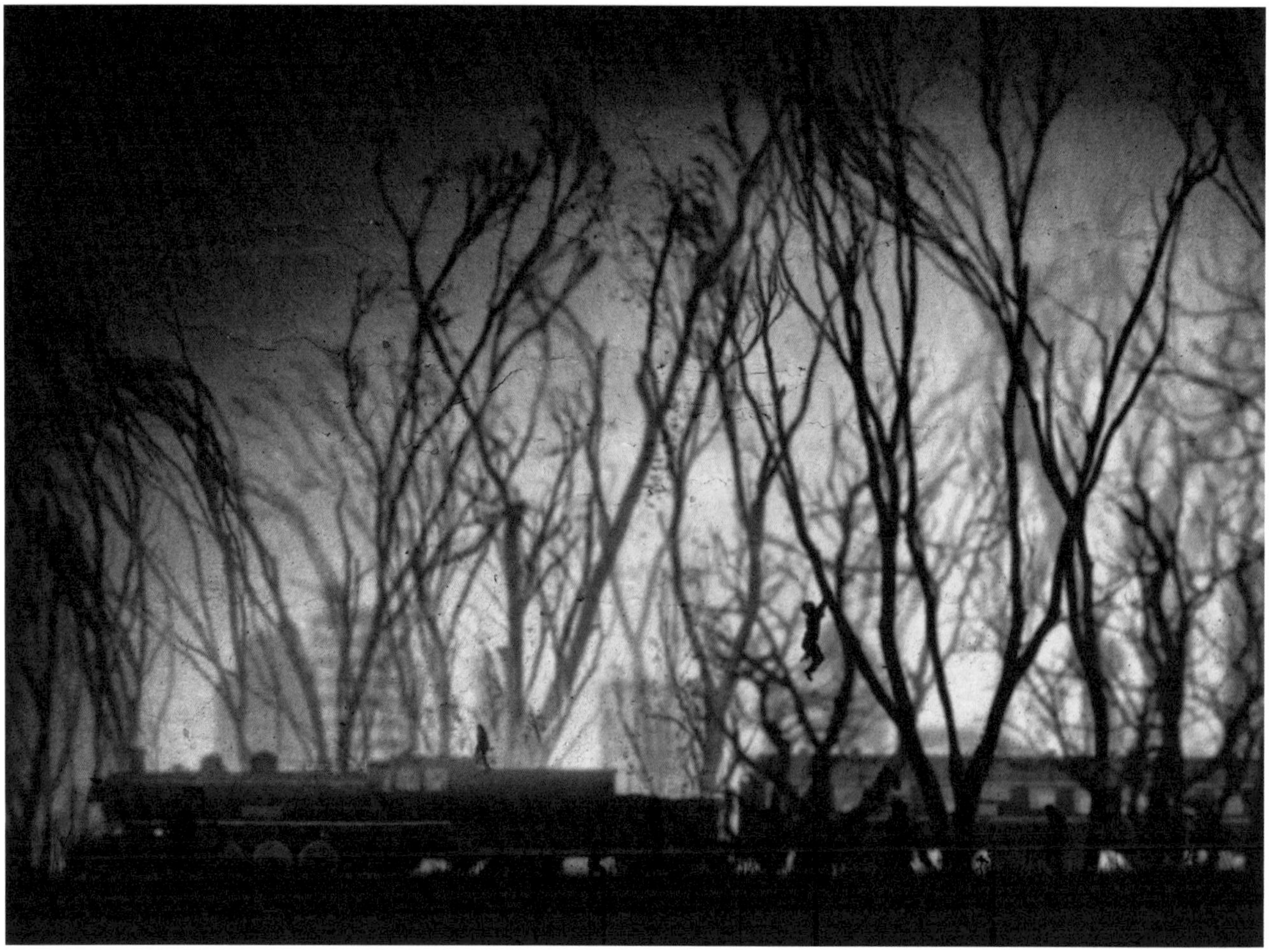

NEW MIRACLE NO.16 /
OUTSKIRTS

SHADOWPLAY SHADE

Pigment print on fibre paper, 50 × 75cm, 2009
Pigment print on fibre paper, 90 × 150cm, 2008–09

Jaap Blonk, sound poet

READING LIGHT
FOR RAOUL HAUSMANN

SPARK RHYTHM ELUCIDATION INTUITIVE

ch ch ch ch ch chtt ch ch ch ch
batt batt batt battatat battchnattbattat
tt uiuz zz z ' —
kkki i i ooooo i i ooooo i u kkiuu kki
hha iiiii u uuuuui ooooo i i i u uuu
oio ou uuachtuachtxex u uuu
oio ou uuachtuachtxex u hhu
hhzz hhzzh hu hhhhhgg gg ggg ggggkhg
bg bbg bbg bbgg bggg e
ccccc c! jjji ji jj
zzzuuuu oooooo o !

ch batt batt battatat battchnattbattat tt
ff ffffrtf fffrrrf frr gggttft—
npokk op oipö
hhhu ui gggz zzgg gz huuuuhu
iop i i i u uuuuui ooooo i u hhu
hhzz hhzzh hu hhhhhgg gg bgggggg—
gghhhgh ggh hhhe gttgt ggt
tt ff ffffrtf fffrrrf frr gggttft—
hhj kkh kh h ggt frr gggttft—
kkki i u uuu oio ou uuachtuachtxex u uu
hgh jjjuu iiiiu jjjuu iiiiu jjjuu hhhe
gttgt ggt tt uiuz zz uuuuhggkzi gkzhi
ggk gr gr gnrr cnrr ch ch ch !
uuuachtachtj hhh h u uuu oio ou
uuachtuachtxex u uu hgh jjjuu hhhe gttgt
ggt frr frff !
cht chchtt chchtt chchtt chatatatta chatt t
att e ccccc c! c! jjji ji jj zzzuuuu
oooooo o !
ch batt batt batt batt batt battatat
battchnattbattat
tt ff ffffrtf fffrrrf frr gggttft—

Bonna Newman, physicist

SOLAR CELLS: THE NEXT GENERATION

ENERGY   SUN   CONVERSION   ELECTRICITY

Light is energy. Every hour, enough light from the sun is absorbed by the surface of the Earth to meet the demands of human consumption for one year. The challenge is to efficiently convert the light energy into the energy form that is most easily transported, stored and used by human civilization: electricity. The direct transformation of light into electricity is known as photovoltaic conversion and underlies the functioning principles of modern solar cells and solar panels. Solar photovoltaics provide less than 1% of the global energy infrastructure today, but are a growing sector. Significant resources are currently dedicated to increasing energy output at lower cost.

Due to the laws of quantum mechanics, each incoming photon can be transformed into one electron. These electrons can then only be extracted from the material at one potential energy. These two principles fundamentally limit the power conversion efficiency, or the ratio of electrical power extracted to incident light energy for a single semiconductor material. Modern, commercially available photovoltaic cells have efficiencies greater than 20%. The best laboratory devices made with a single material are GaAs (gallium arsenide) with 28.8% conversion efficiency[1] and Si (silicon) with 25.6% conversion efficiency.[2] To realize higher efficiencies, a photovoltaic device must operate at the limits of thermodynamics.

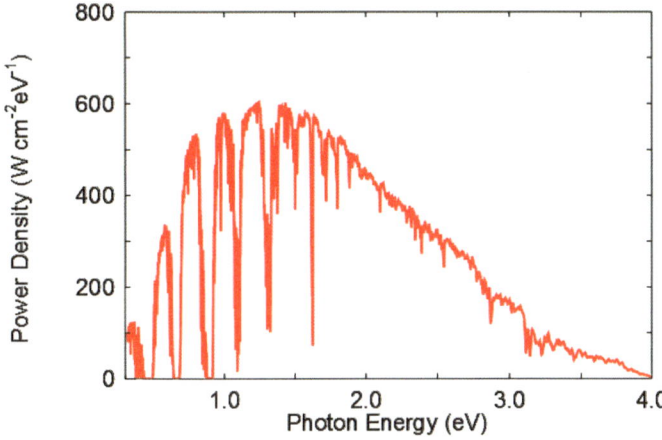

01

Most solar cells today are made from semiconductor material, such as silicon, in which bound electrons are excited into the conduction band by incident photons and then collected in an external circuit. An electron contributes to the current (and energy) in the external circuit as

01   AM 1.5 global spectrum as a function of photon energy.

long as it is in the conduction band. Electrons remain in the conduction band until encountering a missing electron, or hole, and recombining. When electrons and holes recombine, light or heat is emitted due to conservation of energy.

The efficiency of a solar cell is also highly dependent on the incoming spectrum and how well "matched" it is to the absorption spectrum of the semiconductor material. In practice, the incoming spectrum will vary depending on the latitude, atmospheric conditions, pollution, orientation and installation of a solar panel. The standard solar spectrum used in photovoltaics for terrestrial, non-concentration systems is the AM 1.5 (Air Mass 1.5) global spectrum power density, shown in [01] as a function of incoming photon energy.[3] The total power in the incoming spectrum is 1000 W m^{-2} with a peak around 1.4 eV due to the temperature of the sun. Only photons with energy greater than the bandgap of the material are absorbed, therefore light-generated current increases with the decreased bandgap of the material. However, since the electrons are also extracted with maximum potential energy equal to the bandgap energy, there is a fundamental trade-off between decreasing bandgap and the efficiency of the device.

Shockley and Queisser calculated the maximum potential efficiency for a solar device under one sun intensity solar illumination in their seminal paper of 1961, which we outline here.[4] They applied the concept of detailed balance: every conduction band electron created by an absorbed photon must be either extracted into an external circuit or recombine to create heat or light. Mathematically, detailed balance can be expressed as currents such that the light-induced current, I_L, has to be either extracted into an external circuit, I_{ex}, or contribute to recombination, I_R, inside the solar cell:

$$1 \quad I_L = I_{ex} + I_R$$

If the recombination is only radiative, that is, the opposite process of absorption, where a photon is created during the recombination of a hole and an electron, then recombination current is a function of the difference between the photon energy, E, and the electric potential of the charges. In the case of an externally applied bias voltage, V, this relationship is given by a modified Boltzmann distribution:

$$2 \quad I_R = I_{RR}(V) = \frac{2\pi q}{c^2 h^3} \int_{E_{BG}}^{\infty} \frac{E^2}{\exp\left[\dfrac{E - qV}{k_b T_{cell}}\right] - 1} \, dE$$

Here q is the electron charge, c is the speed of light and h is the Plank constant, k_b is Boltzmann's constant, T_{cell} is the temperature of the cell, and E is the emitted photon energy. When there is no external voltage, $V = 0$, there is still a small amount of radiative recombination but it is negligible compared to I_L, and $I_{ex} = I_L$; this current is known as the short-circuit current, I_{SC}. Under short-circuit conditions, no power is created since, $P = I_{ex} \times V$ and $V = 0$.

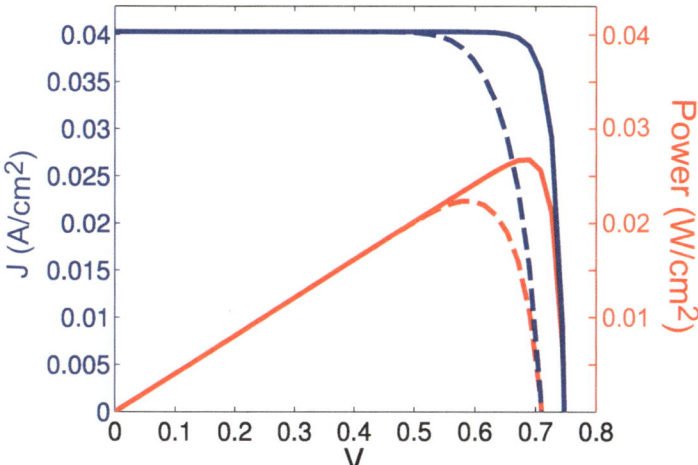

02

To extract power from the solar cell, the bias voltage, V, is increased to a non-zero value. However, as V increases, the radiative recombination increases. This results in a maximum power point at a voltage less than the bandgap energy, as seen in the solid line in [02]. In the case of only radiative recombination, using equations above, we calculate the potential power (efficiency) as a function of bandgap energy for a solar cell with only radiative recombination, as shown in [03]. The maximum efficiency is indicated by the white line with a peak of 33.3% around 1.1 eV.

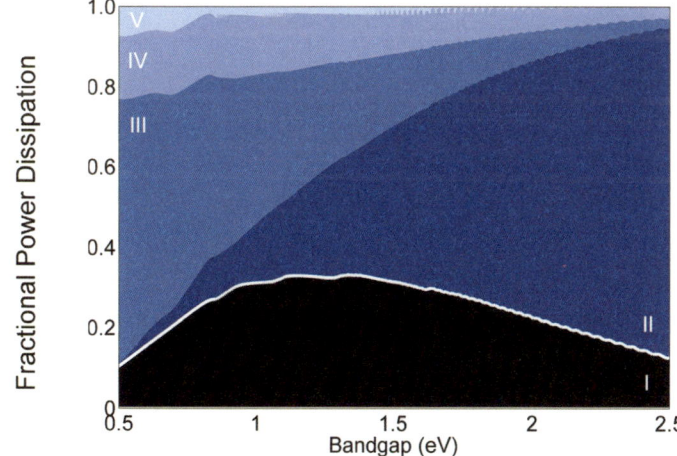

03

02   Current density and power as a function of voltage of a Si solar cell limited by only radiative recombination (solid) and a more realistic cell with non-radiative recombination (dashed line). The maximum power point occurs at a voltage much smaller than the bandgap energy of the material.
03   Fractional power dissipation in a single bandgap solar cell as a function of bandgap under one sun illumination and assuming 100% ERE. Area I is the power that can be extracted in an external circuit. The white line delineates the maximum efficiency under these conditions. Area II is the power lost in non-absorption of photons below the bandgap. Area III represents thermal losses due to electrons excited by high-energy photons that thermalize to the bandgap energy. Area IV indicates losses due to the fundamental thermodynamic limits of energy conversion. And area V is the losses due to radiative recombination. In a device with non-radiative recombination, area V would increase at the expense of area I.

In real materials recombination is not only radiative but also non-radiative. Defects and imperfections in the material, as well as surfaces and dopants, act as recombination centres that do not emit light but instead release heat in the solar cell. Due to these types of recombination, I_R in equation 2 increases, and the potential power and efficiency of a device will decrease, as seen in the dashed line in [02]. The dependence of the bias voltage depends on the exact mechanism of the recombination. Many kinds of recombination can be decreased by applying better surface passivation techniques,[5, 6] or passivating bulk defects;[7, 8] this is a very active area of solar cell research today.

In the best-case scenario, defect-related recombination would be eliminated from the solar cell. But there is another form of non-radiative recombination that is intrinsic to the material, the so-called Auger recombination. In an Auger recombination event, thermal energy from a recombination event is transferred to a second electron already in the conduction band instead of being emitted as light. The fundamental rate limits of Auger recombination in a given material is not fully understood but can be estimated using empirical observations.

The majority of solar cells manufactured today are made from crystalline silicon with a bandgap of 1.12 eV, close to the optimal energy bandgap. However, silicon has an indirect bandgap and is therefore a weak absorber. By reciprocity it is also a weak emitter. Silicon also suffers from relatively high rates of Auger recombination. In spite of these fundamental limitations, the abundance of Si on the earth's surface, as well as inexpensive processing and a well-developed industry, make it an optimal choice for photovoltaic conversion. In the next section, we will specifically look into the fundamental limitations of Si solar cell devices.

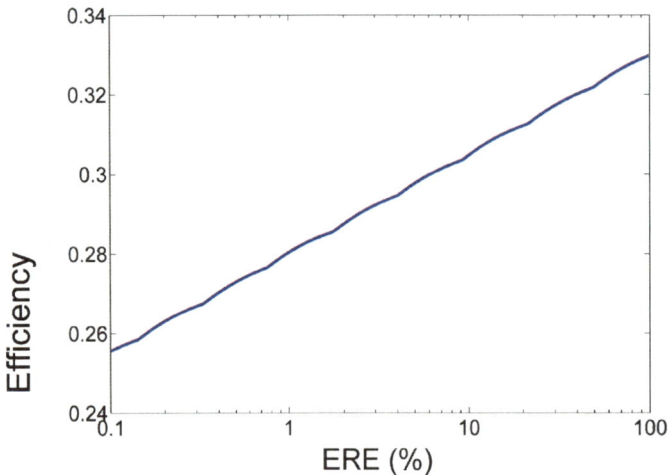

In order to understand the impact of non-radiative recombination, we define a parameter: ERE (external radiative efficiency),[9] which describes the fraction of recombination that is radiative in open-circuit conditions, where $I_R = I_L$. Figure [04] shows the maximum efficiency for a solar cell with a 1.12eV bandgap as a function of ERE. The best Si solar cells today have EREs around 0.5–0.6%,[9] which gives a maximum conversion efficiency around 27.4%. In the best case, where the solar cell is limited only by Auger recombination, silicons will have an ERE of 13.6% with a limiting efficiency of 29.4%.[10]

04   Solar cell conversion efficiency as a function of ERE for a 1.12 eV bandgap device like c-Si.

Therefore, there is a somewhat counterintuitive relationship in solar photovoltaics: the highest-efficiency photovoltaic solar cell will also be a good light emitter. In other words, non-radiative recombination will be minimized and all light that is absorbed would be re-emitted in open-circuit conditions. Since absorption and emission are fundamentally balanced, a strong absorber will also be a strong emitter.

Light-emitting diodes (LEDs) are in many ways the opposite of solar cells; a voltage is applied and light is efficiently emitted. Developments in the field of LEDs has led to a better understanding of these mechanisms, especially with respect to patterning of structures smaller than the wavelength of visible light. Nanostructures offer the ability to specifically tune the optical properties of materials and therefore change the fundamental absorption and emission characteristics. Such developments could have impact on solar photovoltaics as well.[11, 12] For example, nanostructures made of non-absorbing material have been shown to increase the absorption of light in a thin-layer of Si.[13, 14] Nanowires also show promising characteristics, as there is some evidence for increased absorption in these small-scale devices.[15, 16] The use of quantum dots has also been proposed in order to split higher-energy photons into multiple low-energy photons that can be more effectively used in the cells. Industrial solutions for square kilometres of low-cost patterning of nanostructures are necessary for successful application of these ideas in solar photovoltaics.

The above discussion focuses on the limitations and fundamental loss mechanisms for a single bandgap device. As can be seen in [03], increasing ERE only accounts for a small fraction of the lost power. Losses from Area II and Area III due to the mismatch of photon energy with the bandgap of the material limit the ability of single-bandgap devices to reach higher efficiency. Significant research effort is directed at creating tandem or multi-bandgap devices where a system of materials with different bandgaps is used to absorb light from different parts of the spectrum more effectively. These efficiencies have recently surpassed 46%.[17] However, at this point, these devices are expensive and therefore limited to applications where area and volume are a greater concern, such as space or military.

Increasing solar cell conversion efficiencies depends on improvements of not only absorption of light but also highly efficient light emission. Recent developments in the fields of light-emitting diodes, nanotechnology and improved materials offer new ways to improve solar cell efficiencies by fundamentally changing the way solar cells absorb and emit light. In many markets, solar cell efficiencies are already a cost-competitive form of clean energy. With improved technology there remains promise of continued improvements and the potential of solar photovoltaics as a major source of cheap and readily available electricity remains high.

1   B.M. Kayes, H. Nie, R. Twist, S.G. Spruytte, F. Reinhardt, I.C. Kizilyalli, G.S. Higashi (2011), "27.6% conversion efficiency, a new record for single-junction solar cells under 1 sun illumination", *Proceedings of the 37th IEEE Photovoltaic Specialists Conference*

2   Panasonic Press Release (10 April 2014), "Panasonic HIT© solar cell: 25.6% efficiency world record", http://eu-solar.panasonic.net/en

3   ASTM Standard G173–03R12 (2012), *Standard Tables for Reference Solar Spectral Irradiances: Direct Normal and Hemispherical on 37° Tilted Surface*, 14th ed. West Conshocken, PA: ASTM International, www.astm.org

4   W. Shockley, H. J. Queisser, (1961), "Detailed Balance Limit of Efficiency of p-n Junction Solar Cells", *Journal of Applied Physics*, 32:510–519

5   S. Duttagupta, F.J. Ma, B. Hoex, A.G. Aberle (2014), "Excellent surface passivation of heavily doped p+ silicon by low-temperature plasma-deposited SiO x/SiN y dielectric stacks with optimized antireflective performance for solar cell application", *Solar Energy Materials and Solar Cells*

6   G. Dingemans and W.M.M. Kessels (2012), "Status and prospects of Al2O3-based surface passivation schemes for silicon solar cells", *Journal of Vacuum Science & Technology A*, 30, 040802.

7   B.J. Hallam, P.G. Hamer, S.R. Wenham, M.D. Abbott, A. Sugianto, A.M. Wenham, C.E. Chan, X. GuangQi, J. Kraiem, J. Degoulange, R. Einhaus (Jan 2014), "Advanced Bulk Defect Passivation for Silicon Solar Cells", *Photovoltaics, IEEE Journal of*, 4, no.1: 88–95

8   G. Coletti, P. Manshanden, S. Bernardini, P.C.P. Bronsveld, A. Gutjahr, Z. Hu, G. Li (2014), "Removing the effect of striations in n-type silicon solar cells", *Solar Energy Materials and Solar Cells,* 130:647–651

9   M.A. Green (2012), "Radiative Efficiencies of State-of-the-Art Photovoltaic Cells", *Progress in Photovoltaics: Research and Applications*, 20:472–476

10 A. Richter, M. Hermle, S. Glunz (Oct 2013), "Reassessment of the limiting efficiency for crystalline silicon solar cells", *IEEE Journal of Photovoltaics*, 3, 4:1184–1191

11 A. Koenderink, A. Femius, A. Alù, A. Polman (2015), "Nanophotonics: Shrinking light-based technology", *Science,* 348, 6234:516–521

12 A. Polman, H. A. Atwater (2012), "Photonic design principles for ultrahigh-efficiency photovoltaics", *Nature Materials*, 11:174–177

13 P. Spinelli, M.A. Verschuuren, A. Polman (2012), "Broadband omnidirectional antireflection coating based on subwavelength surface Mie resonators", *Nature communications*

14 E.D. Kosten, B.K. Newman, J.V. Lloyd, A. Polman, H. Atwater (2015), "Limiting light escape angle in silicon photovoltaics: ideal and realistic cells", *Photovoltaics, IEEE Journal of,* 5, 1:61–69

15 E. Garnett, Peidong Yang (2010), "Light trapping in silicon nanowire solar cells", *Nanoletters*, 10, 3:1082–1087

16 S. Sandhu, Z. Yu, S. Fan (2014), "Detailed balance analysis and enhancement of open-circuit voltage in single-nanowire solar cells", *Nanoletters*, 14, 2:1011–1015

17 Press Release (1 December 2014), Fraunhofer Institute for Solar Energy Systems

Anton Dekker, artist

STREETLIGHT INTERVENTION

STREETLIGHT INFRASTRUCTURE TRAFFIC POLITICS

Ettore Sottsass, architect, designer

TRAVEL NOTES ON LIGHT
AN EXCERPT

EXPLOSION   INVASION   DARKNESS   GEOMETRY   SHELTER

In the ancient Mediterranean, northeast of Sicily, lies the island called Filicudi. Filicudi is a special island because it has no water springs. And even today there is no light, no electric light. When the sun has set the people use candles or gas cylinders for light, but for the most part they go to bed early. This means that, about nine in the evening, the mountain-island of Filicudi becomes a big, solid-black shadow against sky and sea, likewise dark and phosphorescent. The island plunges into a geological silence, and waits.

Only with full moon the island is bathed in a strange and magic light.

The houses and the terraces glow pale as if it were daytime. The fields are tinged with mother-of-pearl. The caper bushes, the carob-trees, the wild olives and the eucalypti become black holes like caves and no one knows who lives in them. For the island of Filicudi, on certain nights of the year, the moon is a powerful, cold bulb.

What I know is that full moonlight is a cosmic, cold, very frail, soft light that does not illuminate but produces disturbing shadows. I know it digs architecture, rocks and other wonders out of darkness, I know it gives an unexpected, distant, undecipherable, ecstatic description of "place".

I believe innumerable works of architecture have been built on the planet, designed for the light of the full moon and which have vanished perhaps or which I don't know about or which perhaps, aren't known about, at all. There is in any case a feeling that the houses on Filicudi are not designed to receive the light of the full moon. So much for moonlight.

When day begins, early, the island of Filicudi is suffocated by boiling tons of sunlight, as if layers of billions of watts of light have squashed it. In such places everyone escapes. Everyone tries to hide from the light of that murderous lamp.

On Filicudi the houses are designed for survival against the invasion of light by the sun's bulb. On Filicudi the houses are compact and the walls are very thick, built as barriers against the surrounding world. So the rooms have no windows. Or to be precise, they have only one tiny window; a square hole up at the top, almost against the ceiling and through that hole the sun can never get in directly. Only a little dead daylight manages to enter, washing the long black trunks of the ceiling and moving slowly down until it rests on the floor. It comes down more like a grey semi-darkness than light, more to suggest the presence of people, furniture and objects than to illuminate the people, furniture and objects.

Opening the door means exploding a sudden bomb of light into the room. It means being blinded, opening the diaphragm too wide, overexposing the vision of reality. Also it means seeing and being seen; it means over-elongating the shadows of even the smallest flies and beetles; it means giving the weight of stone to the folds of sheet on the unmade bed. For this reason the doors of rooms, which are glazed, are in their turn covered by small shutters that can be opened without even opening the door, to ensure a still more meticulous filter against the danger of light.

This idea of the shade and a solid, total darkness as the sole place of survival, as a more than physical, at times metaphysical balm, can be recognized in the remains of some ancient Egyptian temples. You only have to venture into those immense blocks of architecture, filled with countless gigantic columns, to feel that darkness has become the absolute mistress of existence, the place of terror and protection, a final image of reality.

In India too, I have seen temples that were built simply to produce a compact, sudden, solid, metaphysical darkness.

I believe innumerable works of architecture, houses, palaces and temples have been built to the theme of creating shade – a total, mysterious shade; a shade that immediately becomes a visible and tangible metaphor for the impenetrable structure of existence.

The ancient people of those countries had invented an incredible number of things for governing light. They had captured it in precise forms, diagrams and memoranda, in recognizable and orderly signs for performing the ritual of life.

The result is that those vast places, those deserts or islands or volcanoes without water or shade, are full of architecture built around diagrams of light and shade, where the definition of environment is entrusted to light so that the daily lives of its inhabitants can be kept in a state of clarity, concentration and permanent awareness. There are no other explanations for the architectural organization of Norman cloisters, Arabian courtyards or Egyptian houses, Greek peristyles or whatever. They are places to be reached slowly, just as it takes time to reach the far end of the Madrasa Qala'un, the mosque, tomb and maristan in Cairo. To reach it you have to go through successive baths of controlled lights, from the dazzling sweat of the street, through successive purifications until isolation, concentration and final serenity are accomplished in the fixed and suspended half-pale shadowy light of the courtyard, the sacred, silent heart of the architectural cosmos.

If you go into the Madrasa Qala'un, the light switches suddenly from the white dust of the street sun to the dense, heavy, vulgar darkness of a corridor with a black and cool paved floor, a low corridor with no windows and no air. In that corridor you can do nothing but abandon everything outside. You begin a different existence because darkness at those levels obliterates everything. You start a new journey.

Then you continue through a large, empty and tall, grey room with an opaque earthen floor, lit by small windows high up beneath the ceiling, windows with big, coloured panes. Space here begins to be suspended. What you have absentmindedly noticed each day on the ground doesn't seem to matter any more. You tread lightly now, floating on air, filled with colours. And from now on the space to be discovered grows more and more sophisticated.

The route continues between two very high, bare walls of a narrow corridor with no ceiling. You can see the distant sky, but by the time any light gets to the bottom of that well it is tired and obscure, worn out by the innumerable reflections, refractions and interruptions on the way down.

Perhaps the ordeal of sudden darkness is followed by a return to a new possible measure of existence, perhaps to a new possible idea of freedom. Now that the senses are attuned to more sensitive channels, they may perhaps have a fresh, profound consciousness of freedom.

Next, you pass through a very large square hall with a marble floor. The hall is closed, illuminated by gigantic windows with gratings that inject billions of grains of light into the darkness, suspended like seaweed drifting in dark water.

At last you come on to the polished pavement of the immense, open courtyard, with its very high walls enclosing in total silence a vast cube of light without directions or colour. A cube of grey, faded light, as abstract as thought, a limbo-light with neither reflections nor shadows. Where things become simply thought, where thoughts at last unfold like childish paper flowers in the slow geometries of consciousness.

NOTE
Qala'un (or Qalawun) was a sultan of Egypt (1279–90). The eponymous complex contains a mosque, tomb and hospital.

Ruxandra Mitache, artist

UNTITLED GARDEN

EXPOSURE | PLAY | GARDEN | ATMOSPHERIC | HYPNOTIC | ENCHANTMENT | TRANSPARENCY

I created a small garden of light using photo paper, which was exposed to natural light filtered by a painted transparent material and then edited digitally.

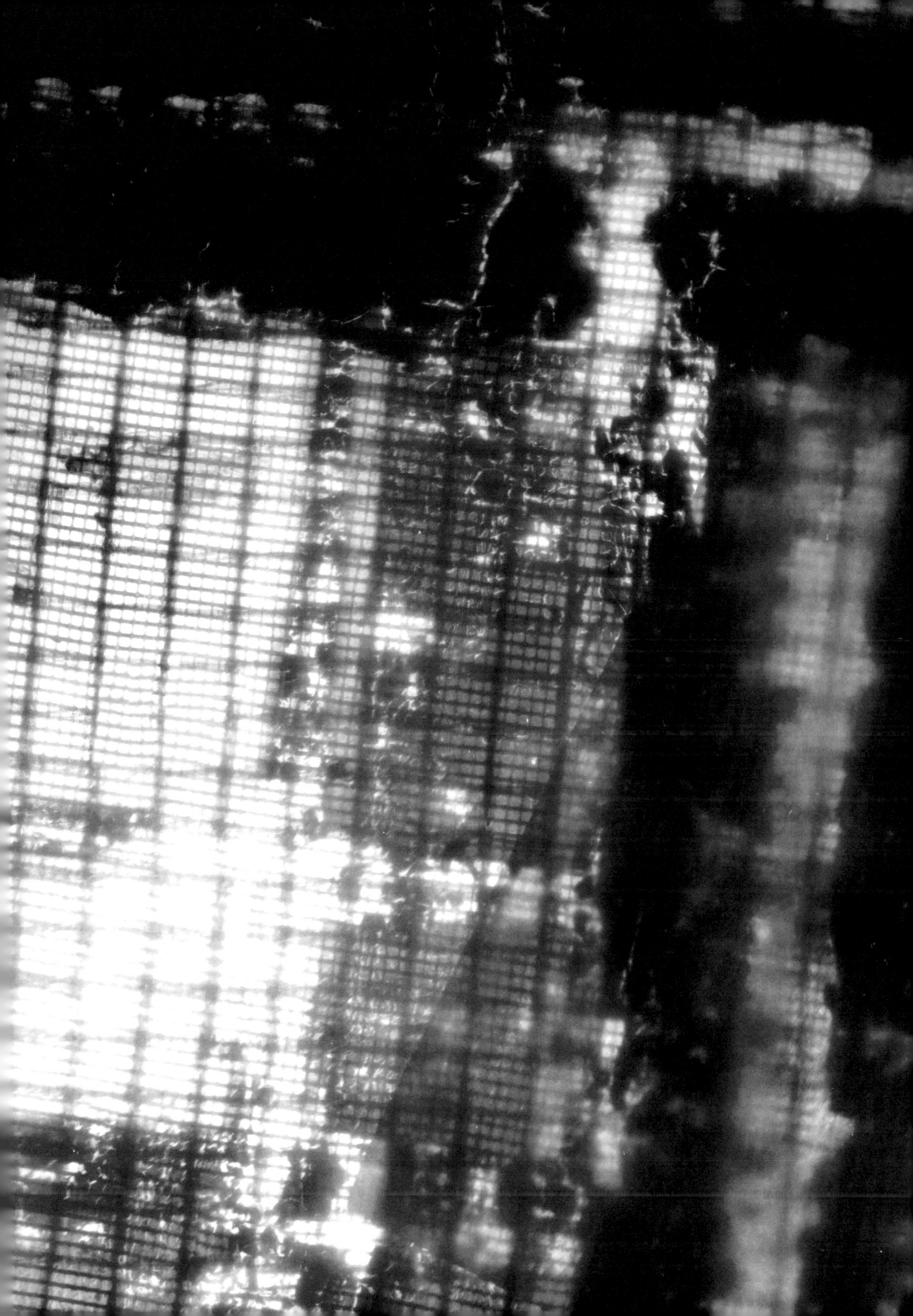

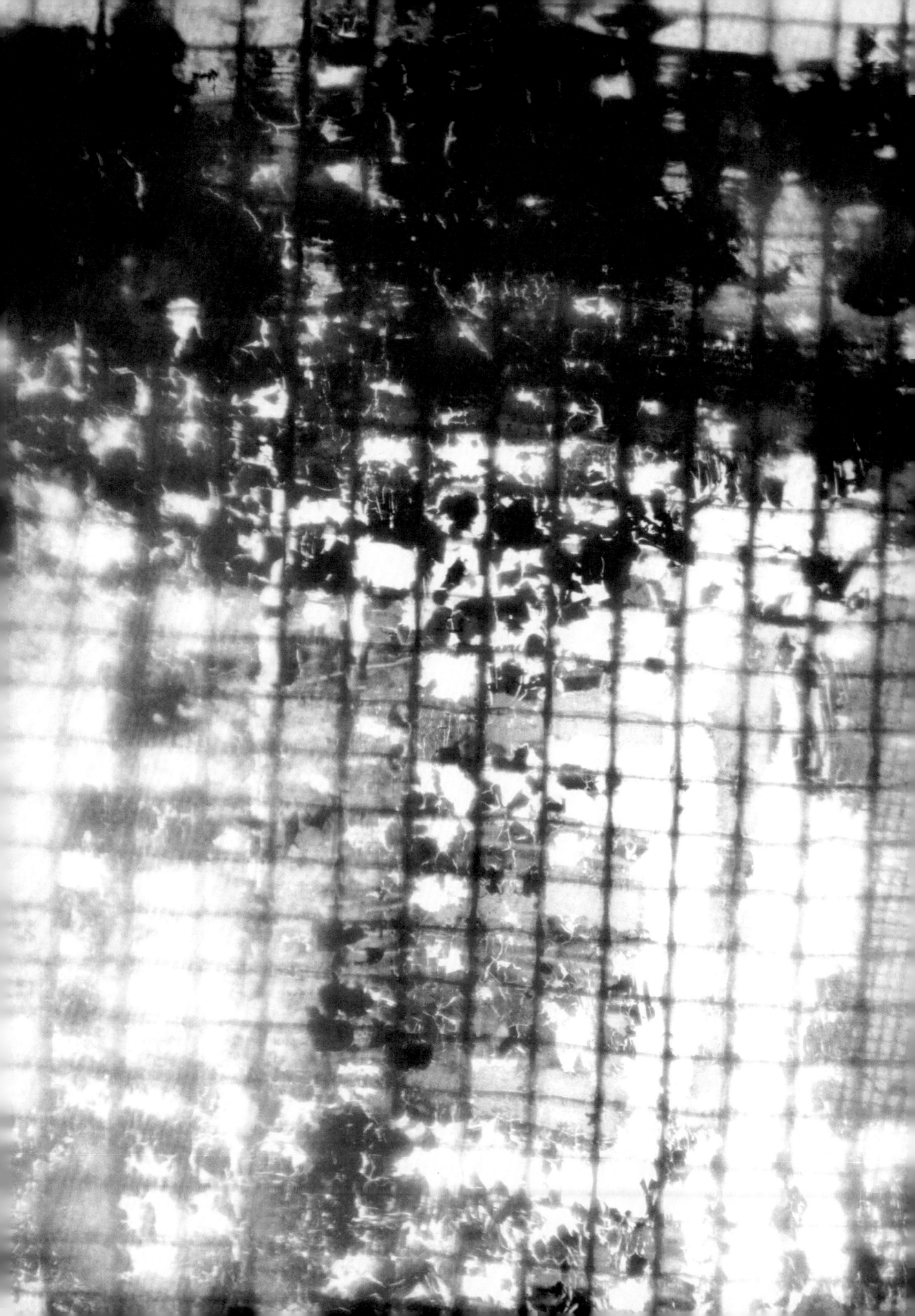

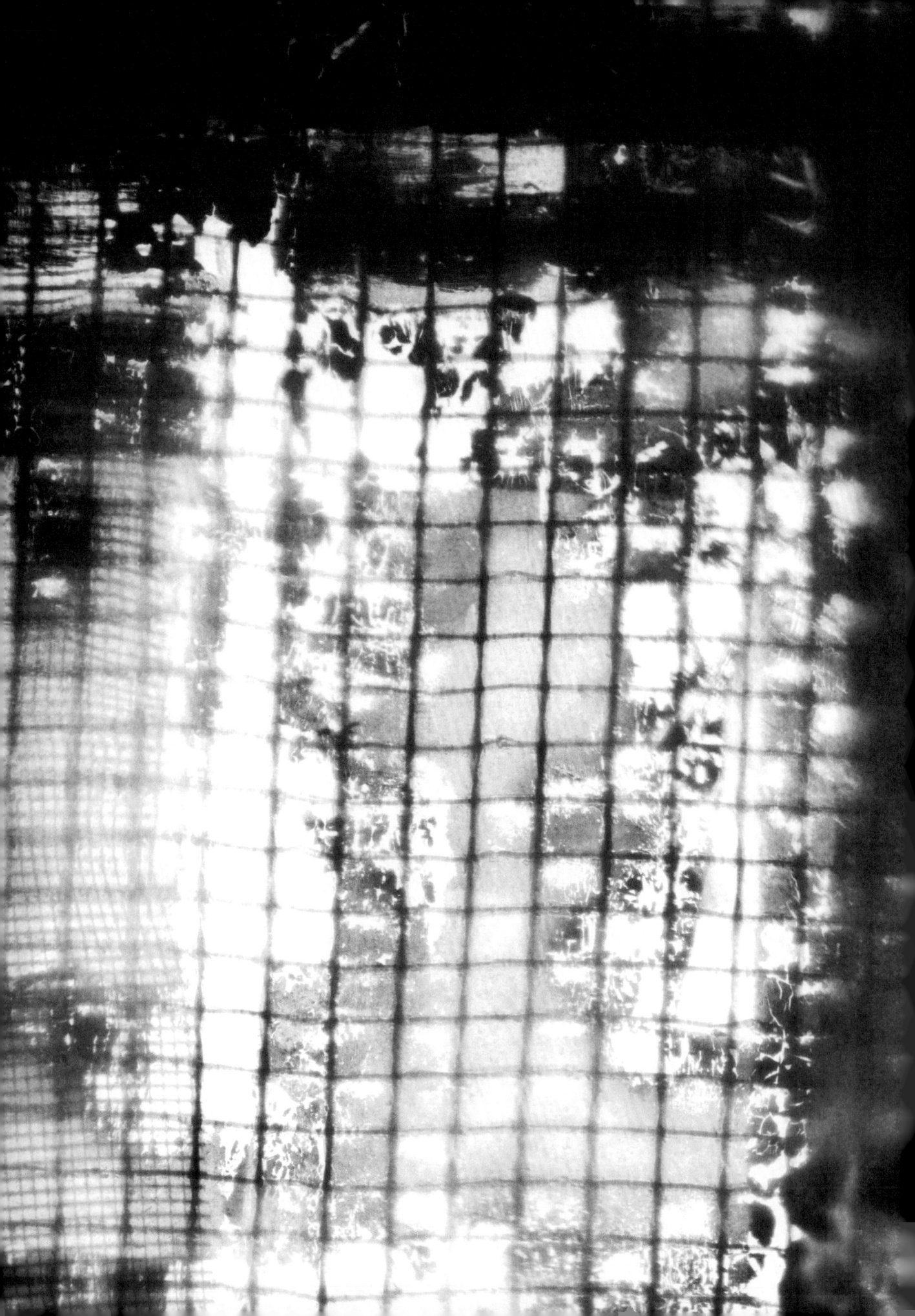

HOW THE MIND WORKS
AN EXCERPT

ILLUSION   CURIOSITY   MYSTERY

Visual illusions have always captivated people. Simple illusions made up of parallel lines that seem to converge and congruent lines that look unequal have long appeared in cereal-box reading material, crackerjack prizes, children's museums, and psychology courses. Their fascination is obvious. "Who are you going to believe, me or your own eyes?" says Groucho Marx to Margaret Dumont, playing on our faith that vision is a certain route to knowledge. As the sayings go: I call them as I see them; Seeing is believing; We have an eyewitness; I saw it with my own eyes. But if a devilish display can make us see things that aren't there, how can we trust our eyes at other times?

Illusions are no mere curiosities; they set the intellectual agenda for centuries of Western thought. Sceptical philosophy, as old as philosophy itself, impugns our ability to know *anything* by rubbing our faces in illusions: the oar in the water that appears bent, the round tower that from a distance looks flat, the cold finger that perceives tepid water as hot while the hot finger perceives it as cold. Many of the great ideas of the Enlightenment were escape hatches from the depressing conclusions Sceptical philosophers drew from illusions. We can know by faith, we can know by science, we can know by reason, we can know that we think and therefore that we are.

Perception scientists take a lighter view. Vision may not work all the time, but we should marvel that it works at all. Most of the time we don't bump into walls, bite into plastic fruit, or fail to recognize our mothers. The robot challenge shows that this is no mean feat. The medieval philosophers were wrong when they thought that objects conveniently spray tiny copies of themselves in all directions and the eye captures a few and grasps their shape directly. We can imagine a science-fiction creature that embraces an object with calipers, prods it with probes and dipsticks, makes rubber moulds, drills core samples, and snips off bits for biopsies. But real organisms don't have these luxuries.

When they apprehend the world by sight, they have to use the splash of light reflected off its objects, projected as a two-dimensional kaleidoscope of throbbing, heaving streaks on each retina. The brain somehow analyzes the moving collages and arrives at an impressively accurate sense of the objects out there that gave rise to them. The accuracy is impressive because the problems the brain is solving are literally unsolvable. Inverse optics, the deduction of an object's shape and substance from its projection, is an "ill-posed problem", a problem that, as stated, has no unique solution. An elliptical shape on the retina could have come from an oval viewed head-on or a circle viewed at a slant. A patch of grey could have come from a snowball in the shade or a lump of coal in the sun. Vision has evolved to convert these ill-posed problems into solvable ones by adding premises: assumptions about how the world we evolved in is, on average, put together. For example, the human visual system "assumes" that matter is cohesive, surfaces are uniformly coloured, and objects don't go out of their way to line up in confusing arrangements. When the current world resembles the average ancestral environment, we see the world as it is. When we land in an exotic world where the assumptions are violated – because of a chain of unlucky coincidences or because a sneaky psychologist concocted the world to violate the assumptions – we fall prey to an illusion. That is why psychologists are obsessed with illusions. They unmask the assumptions that natural selection installed to allow us to solve unsolvable problems and know, much of the time, what is out there.

Echo Ho, video artist

ARRIVE BEFORE DAWN

OVERCAST ESTRANGEMENT MESMERIZING

Stills from video installation triptych.

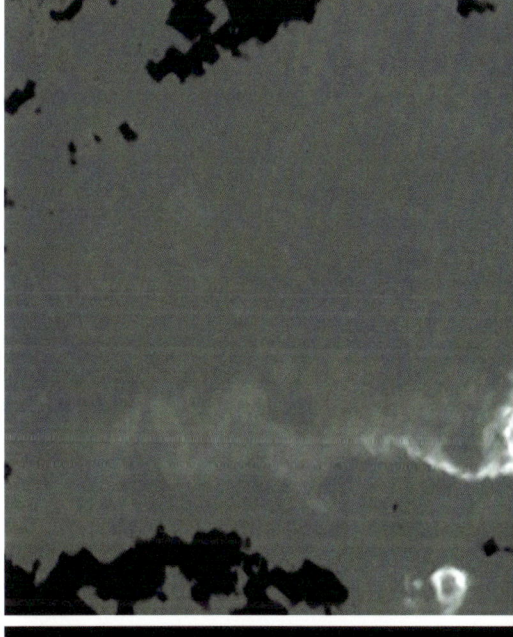

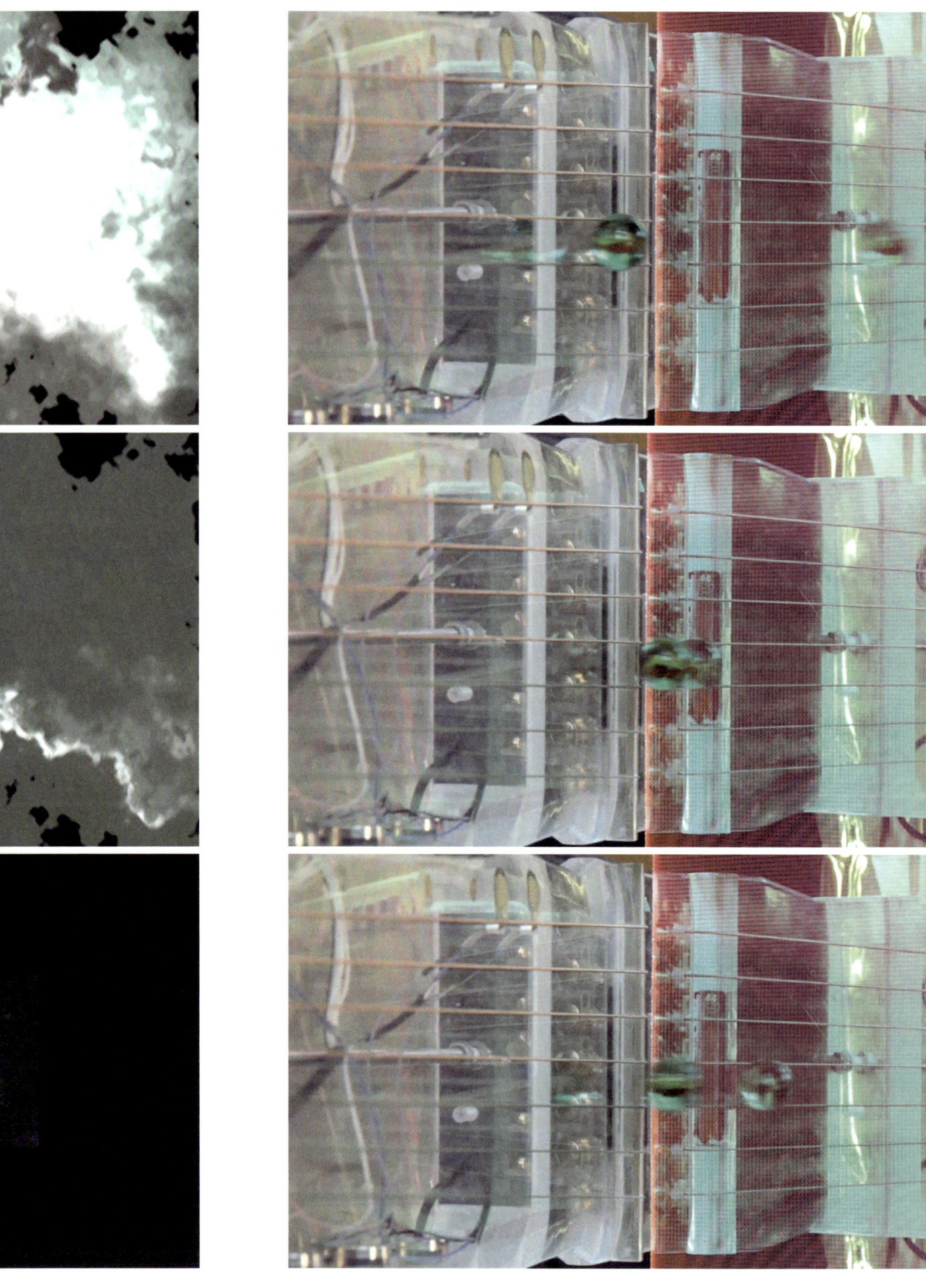

Arthur Zajonc, physicist

REFLECTIONS

LUCIFER AND RA | BEAUTY | ELUSIVE | IMAGINATION | INSIGHT | RELIGION | INNER LIGHT

I'll tell you how the sun rose,–
A ribbon at a time.
[Emily Dickinson, 1924]

Every insight into light arises through relationship. Every finding concerning light therefore reflects as much about the investigator as about light itself. The decades I've spent inquiring into light have convinced me that the subject is inexhaustible. With each new insight, a new pleasure dawns. May these passages give you a glimmer of the pleasure that I've experienced.

I first sought to understand light by means of laboratory research in quantum optics. In laser experiments performed at institutes in Boulder, Amherst, Paris, Hannover and Munich, I studied light and the way it touches matter. The more I learned of the quantum theory of light, theoretically and experimentally, the more wonderful light seemed. Even armed with such sophisticated theories, I have no sense of closure regarding our knowledge of light. Far from it, light remains as fundamentally mysterious as ever.

Once I understood that, for all the power, precision and beauty of quantum optics, we still do not know what light is, I got excited.

The first thing I discovered was that light has gathered around it innumerable artistic associations of extraordinary beauty. Light touches all aspects of our being, revealing a part of itself in each encounter. What will light look like tomorrow? From the birth of the very idea of light, light will have sighted whole kingdoms and nourished prairie, tree and flower. In the mingling of nature and mind arises an understanding of the life of light. The outer light of nature, also the inner light of the mind. I have grown convinced that the two are inseparable.

I am the one who openeth his eyes, and there is light;
When his eyes close, darkness falleth.
[Ra speaking; from the Turin papyrus, 1300 BCE]

Two eyes looked down on the civilization of the Nile, the "two eyes of Horus", the sun and moon. No more significant symbol existed in ancient Egypt than the eye of the sun-god Ra. His eye – the sun – was creative, his vision was life itself. It was said that mankind arose from the tears of his eye. In Egyptian the very words for tears and men sounded similar. The gaze of Ra was the light of day. For men and women of that civilization, to stand within daylight was to stand in the sight of their sun-god. The power of vision to illuminate the world was universalized, projected on to the grandest scale, becoming the brightness of day. The gaze of God was light. *Light was God seeing.*

How you have fallen from heaven, bright son of the morning,
felled to the earth ...
[Isaiah 14:12–15]

The fall of man was inseparable from the fall of the bearer of light, Lucifer. As angelic light was cast down on to the earth, man and woman ate the fruit of the tree of knowledge, and so were cast out to labour in the lands east of Eden. The expulsion and fall of Lucifer, and the accompanying fall of man, finds its counterpoise in the incarnation of Christ – luminous being of the sun. According to the Gospel of John, Christ is "the light of the world", a light that shines into the darkness. Christ was to be understood as the "real light" – the true spiritual, consummately angelic light of the world.

Then as we walked, there was
a heaped-up cloud ahead that changed into a tepee,
and a rainbow was the open door of it.
[*Black Elk Speaks: being the life story of a holy man
of the Oglala,* Black Elk, 1932]

Look at the dew in the morning sun. Each droplet sparkles like a gemstone. Find one that shines brightly, and slowly move your head up and down. The glistening colours you see always pass through the same ordered sequence of the rainbow: red, yellow, green, violet. Place these dewdrops in the sky as rain, and they continue to glisten, each with a single colour. Move your head again and new droplets shine red while the old ones turn yellow. Geometry is everything in the figure of light. Colourless in itself, each droplet sparkles, a diamond of coloured light standing between the eye of Horus and the eye of man.

All things linked are, that thou canst not stir a flower,
without troubling a star.
["The Mistress of Vision", Francis Thompson, 1859–1907]

If one day we awoke and truly saw the world this way, the ramifications would shake our psyche to its foundations. Assuming we remain sane, the relationship of I to thou, of individual to planet, of my actions to yours, would be revolutionized. Edward Lorenz's theory of nonlinear systems delivered the "butterfly effect". The flight of a butterfly in Rio de Janeiro can, in fact, change the weather of Japan. If chaos dynamics have shown us the extreme delicacy of our world, quantum physics reveals its deep intimacy through the phenomenon of quantum entanglement.

When I try to imagine light without a particular colour or direction of propagation, I understand the struggles of medieval theologians and artists as they sought to portray God, and appreciate their choice of light as an attribute of the divine. God's existence was not at issue, but his features were. Anything definite one said about them, any image that depicted Him, was of necessity partial and potentially misleading. Even such a simple thing as the location of God, where God exists, was full of dangerous contentions. Amazingly enough, light, too, suffers from a special difficulty with regard to location. We have seen its other attributes entwined in subtle ways, but what of the simple sense of place?

Love is not consolation, it is light.
[Simone Weil, 1954]

Everything must have location, a place where it is. What then is the place of light? One would suppose that, with the photon concept of light, a straightforward response would be forthcoming; but quantum theory and experiment again conspire to make the place of light completely elusive. Many times over the last 60 years, the quantum theory of light has been searched for a way to give a position to light, and it has consistently denied what it otherwise provides so readily for massive particles. Why is light so recalcitrant about declaring its position? The answer seems connected to the transverse nature of the electromagnetic field.

Every assumption we make about light, assumptions common to us from daily life, leads to errors. When it comes to light, we cross into another domain, and must learn to leave behind what we hold dear from the past, and cleave only to the archetypal phenomena of light at every level, down to the quantum. Particles, waves, location ... all should be left, like soiled sandals, at the threshold of the temple.

Is there a light in darkness? Is the night empty, void and dead, or does it, too, offer more than appearances suggest?

> Our whole business in this life is to restore to health the eye of the heart whereby God may be seen.
> [Sermon 88.5.5, St Augustine, 354–430]

To the medieval imagination two worlds, the physical and the spiritual, were one within the sanctuary of the cathedral; and light played a leading role in that unitary imagination of God in man's domain.

> Finally, I must tell you that as a painter I am becoming more clear-sighted before nature.
> [Letter to his son, Paul Cézanne, 1906]

The eye becomes concentric, aligned with nature, through the artist's ceaseless action of looking and working, of struggling to see clearly a single gesture of nature's infinitely varied repertoire, and then to paint it. In guiding the hand across a canvas, one fashions and refines fresh senses, new capacities of mind suited to seeing, that which until then had eluded the eye. In the end, one "gets to the heart of what is before us". Like the alchemist, whose outer actions were but an image of an inner transformation, the artist, in creating outwardly, simultaneously accomplishes an equally precious inner work: *clearsight*.

The philosopher Schopenhauer once recorded a remarkable conversation from 1813 between Goethe and himself concerning light.

Schopenhauer sensibly suggested that light is a purely subjective, psychological phenomenon, and that without sight, light could not be said to exist. Goethe responded vehemently, as Schopenhauer describes it:

> "What," he [Goethe] once said to me, staring at me with his Jupiter-like eyes. "Light should only exist in as much as it is seen? No! You would not exist if the light did not see you!"
> [Goethes Gespräche, 1813]

If every culture and period have thought so variously about light, and if quantum theory has stripped light of its naively imagined attributes, then what of certainty is left to the nature of light at the end of this evolutionary drama? Everything. The true artist, monk and scientist are not searching to grasp knowledge as object, but rather as event. The moment of critical importance is the moment pointed to by Goethe, the moment of *aperçu*, or insight. For millennia one can see the sun rise and never notice the rotation of the earth. One can throw a thousand rocks and never see their parabolic flight. We can wake each morning for 60 years to the glow of the dawn and never see light. Why? Because one rushes past the immediate offerings of the senses to what we suppose to be the hidden, enduring, primary objects of reality. The habits of our culture, the dogmas of our education constrain our sight. Atoms become immortal gods, photons their stern messengers.

> All the fifty years of conscious brooding have brought me no closer to the answer to the question, "What are light quanta?"
> Of course today every rascal thinks he knows the answer,
> but he is deluding himself.
> [Albert Einstein, 1951]

Over millennia, cultures have embraced and discarded countless images of light. Within a single lifetime, likewise, we have lived within and shed successive understandings of light. Through research, artistic praxis, and quiet contemplation, light's elusive being constantly recreates itself in our mind's eye, offering fresh epiphanies to every generation. When seen with a thousand eyes, light will finally rest with us in the haven we have made.

Every imagination of light has been held within a larger cultural imagination of man and world. Are we now at the watershed to a new one, and could it possibly support a true ecology of human, animal, plant and mineral communities? In recent decades, light has taken on a new and subtle figure; one can only hope it is but a symptom of a larger evolutionary change in the structure of our imagination that supports an ecological consciousness.

Seeing light is a metaphor for seeing the invisible in the visible, for detecting the fragile imaginal garment that holds our planet and all existence together. Once we have learned to see light, surely everything will follow.

Eleni Kalorkoti, illustrator / Astrid Alben, poet

MOONLIGHT IN A BOX

DREAM | BIRDSONG | MOONLIGHT | IMAGINATION | BEGINNING

When the moon hides behind
 the curtains
I scoop the moonlight up with a box
 but the box won't hold her.

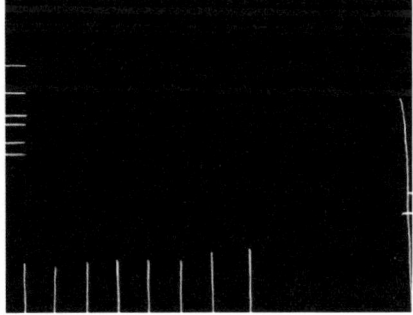

When the moon blinks in the rush
 of the stream
the scientist puffs he's not quick-witted
 enough to hold her.

When the ditch is filled with birdsong in
a silver seam—
Where are you moon?
Where have you gone without me?

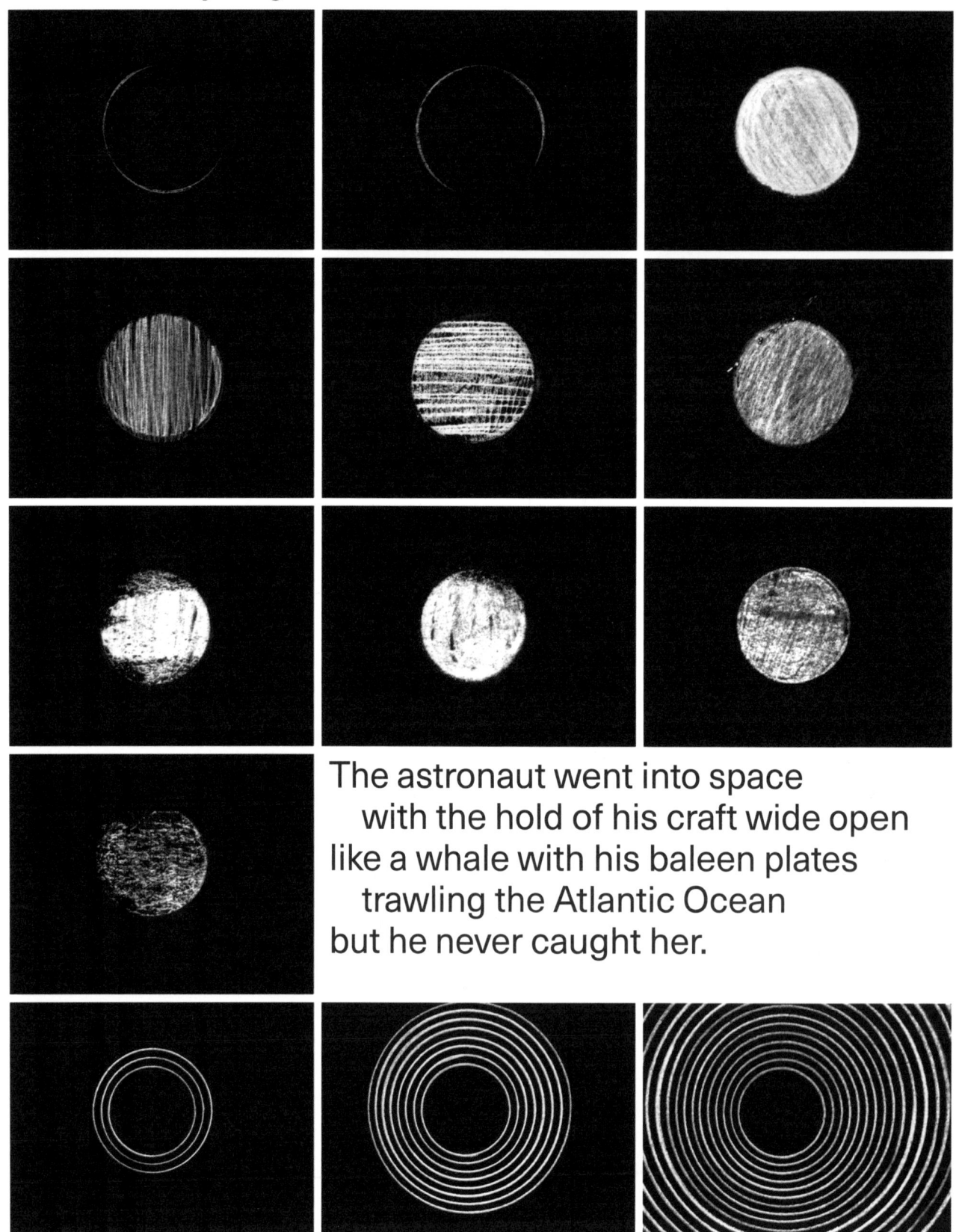

The astronaut went into space
   with the hold of his craft wide open
like a whale with his baleen plates
   trawling the Atlantic Ocean
but he never caught her.

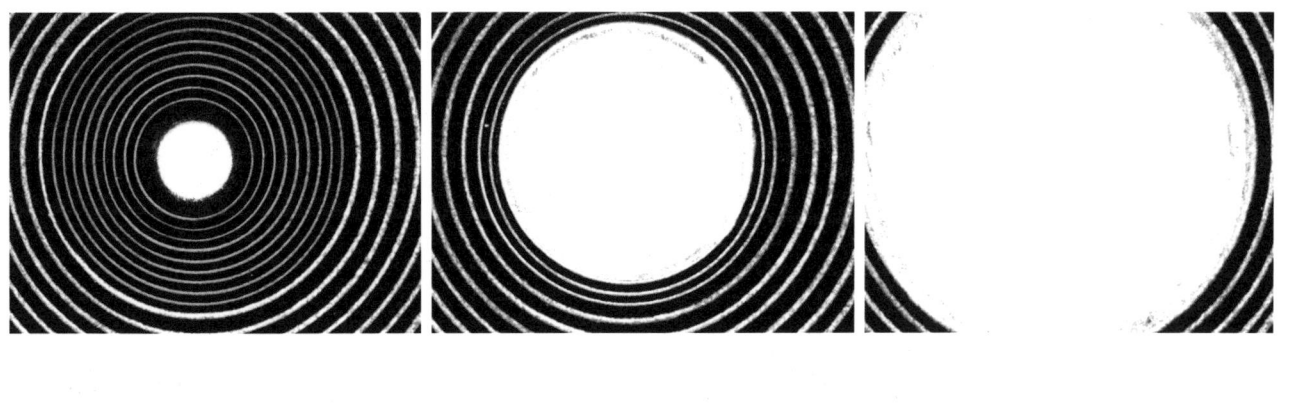

At dawn I looked in my cardboard box.
Inside was everything
     I had believed was outside—

dark, empty and only just beginning.

Contributors

Adam Fuss (UK, 1961) is a photographer based in New York City. His work is distinctive for its contemporary reinterpretation of photography's earliest techniques, particularly the daguerreotype and the camera-less photogram. His work is represented in many American and international collections including MOMA and the Victoria and Albert Museum.

Anna Wirz-Justice (New Zealand/Switzerland, 1940) is emeritus Professor at the University of Basel Psychiatric Clinic, where she founded the Centre for Chronobiology, specialized in research on human circadian rhythms, sleep, mood and performance, as well as diagnosis and treatment of circadian and seasonal disorders. She introduced light therapy for winter depression to Europe and extended the application of light to other illnesses, from non-seasonal affective disorder to Alzheimer's disease and sleep disturbances in medicine. Her present interest is the translation of research findings into architectural use of (day)light to influence wellbeing and health.

Annica Cuppetelli (USA, 1977) and Cristobal Mendoza (Venezuela, 1976) began collaborating as Cuppetelli • Mendoza in 2010 and are based in Detroit. They create installations and objects that combine physical elements with digital technologies, composing sensual, immersive and dynamic experiences. Their work is exhibited internationally, including at the Bienal de Video y Artes Mediales in Chile and Denver Art Museum.

Anton Dekker (Netherlands, 1966–2015) was an artist who primarily developed his performance work and built installations in public spaces. Dekker was interested in what is often not visible and what is about to change: the passage of time, the past, present and future of objects, buildings, landscapes and people. His working methods and processes were often public as he considered them as an integral part of the work.

Arthur Zajonc (USA, 1949) is President of the Mind & Life Institute in Hadley, Massachusetts and emeritus Professor of Physics at Amherst College. His research has included studies in electron-atom physics, parity violation in atoms, quantum optics, the experimental foundations of quantum physics, and the relationship between science, the humanities and the contemplative traditions. He has written extensively on Goethe's science work and is author of the book *Catching the Light*.

Arturo Fuentes (Mexico, 1975) is a composer who studied under Franco Donatoni and Horacio Vaggione. Fuentes composes instrumental and electronic music as well as music for theatre and film. Fuentes writes about his creative process: "I select a sonorous line and walk with it. The breaks in the line, changes in speed, suspended moments, etc – all of these things lead to the emergence of a sonorous form and musical context." He lives and works in Vienna and Innsbruck.

Bianca van der Zande (The Netherlands, 1970) is a Senior Scientist at Philips Research, where she leads research in the area of human-centric lighting, which studies how light affects the human body. She translates scientific knowledge about the biological effects of light into new lighting systems for a broad range of application areas, including nursing homes, schools, offices and homes. Van der Zande holds a doctorate in Physical and Colloid Chemistry from the University of Utrecht.

Royal Philips is a diversified health and wellbeing company, focused on improving people's lives through meaningful innovation in the areas of Healthcare, Consumer Lifestyle and Lighting.

Bill Carlton (England, d. 2012) worked for 30 years as a designer and chief lighting engineer. He toured the world training and lecturing in lighting. He was responsible for the interior lighting of Salisbury Cathedral and the floodlighting of Westminster Abbey, Bath Abbey and the Roman Remains Museum, among many others. Bill's collection has been exhibited at the Science Museum and at the Institute for Engineering and Technology in London in 1975 and 1979.

Bonna Newman (USA) received her PhD in experimental physics from MIT in 2008. She received the Claire Boothe Luce Postdoctoral Fellowship from the MIT Energy Initiative for research in fundamental material properties for photovoltaic applications. She focuses on developing high efficiency solar cells on less than 20 micrometres of high-quality crystalline silicon material. She also founded SunFly Consulting Inc., which assists various companies in developing improved and integrated processes for silicon solar cells as well as doing photovoltaic market research. In addition to her research, she is actively involved in encouraging women and minorities in science.

Bradley Pearce (England, 1988) is a neuropsychologist fascinated with how perception remains constant while the image at the eye changes. He is currently a researcher in Professor Hurlbert's laboratory, working on the EU Framework7 project, HI-LED. He also makes software that utilizes principles of human vision and writes computer games.

Dinie Besems (The Netherlands, 1966) is a designer known for her radical and conceptual approach inspired by natural processes and phenomena. Her work ranges from jewellery to public art, and from pure digital designs (often made in 3D) to craft designs. Besems' work is characterized by exploring – and exceeding – boundaries and thus changing meaning and value. Her work appears in the collection of several prominent museums.

Echo Ho (China, 1973) is a Cologne-based artist, whose artworks cover installations, objects, situational interventions, live performances and audiovisual compositions. Ho re-invented the oldest traditional Chinese string instrument, the gu qin, as a "slow qin", taking audiences on an unexpected sonic journey by playing "slow qin" tunes with field-recorded-like sounds, and merging voice with electronic noise and experimental beats.

Eleni Kalorkoti (Scotland, 1986) is an illustrator. She is known for her minimalist, graphic yet playful illustrations and works for clients such as the *New York Times*, The Victoria and Albert Museum and Random House. She now lives in South London where she splits her time between client work, self-publishing projects and dog walks.
Astrid Alben's (The Netherlands, 1973) most recent poetry collection *Ai! Ai! Pianissimo* was published in 2011. Alben has been described as "a new and original voice in English poetry, serious and uncompromising" (*TLS*). Her poems, essays, translations and reviews are widely published, including in the *Best of British Poetry Anthology 2015*. Her poetry is translated into Romanian, Dutch, Slovenian, Chinese and Japanese. Her next collection, *Plainspeak*, an alter-ego-thinker-out-louder book, will be ready this year.

Ettore Sottsass (Austria, 1917–2007) was an Italian architect and designer. As one of the founders of the Memphis Group (1981) he made limited-production creations of unusual objects and functional designs that featured plastic laminated surfaces, bright colours and bold patterns.

François Morellet, (France, 1926) is one of the major figures of post-war geometric abstraction and a forerunner of minimal art.

Guy Ropars (France, 1957) is Associate Professor at the University of Rennes. His dissertation dealt with the dynamics of eigenstates and vectorial bistabilities in lasers. He is currently involved in theoretical and experimental studies of stochastic effects on polarization switchings in vectorial bistable lasers and in human vision.

Albert Le Floch (France, 1939) is Professor of Physics at the University of Rennes and Head of its Laser Physics Laboratory. His range of interests span from polarization effects in lasers and human vision, electromagnetism at interfaces including Newton-Wigner delays to chiralities in molecular systems.

Hamid Ismailov (Kyrgyzstan, 1954) was born into a deeply religious Uzbek family of Mullahs and Khodjas, many of whom lost their lives during the Stalin-era persecution. Though he could have become a high-flying Soviet or post-Soviet apparatchik, instead his fate led him to become a dissident writer and poet residing in the West. He was the BBC World Service's first Writer in Residence. His work is informed by the Russian classics, Sufi parables and Western postmodernist thought, and his unique intercultural experience draws the reader into his world.

Hans Kristian Eriksen (Norway, 1977) is Professor of Astronomy at the University of Oslo. His speciality is statistical analysis of the Cosmic Microwave Background. In 2010 he won an ERC Starting Grant to study the large-scale statistical properties of the universe, and in 2011 he was awarded a Leverhulme visiting professorship at Oxford University. His current research focuses on the Planck experiment, ESA's cosmology flagship satellite mission.

Iván Navarro (Chile, 1972) is a light artist based in New York, known for his sociopolitically charged sculptures of neon, fluorescent and incandescent light. In his work, light, form and power come together both as the material base for his sculptures and installations, and as an underlying subtext exploring authority, punishment and control. His work is presented at numerous international exhibitions and is held in public and private collections. He currently lives and works in Brooklyn.

Jaap Blonk (The Netherlands, 1953) is a self-taught sound poet, composer and performer. He is currently on a world tour with his soundscapes and sound poetry. A comprehensive collection of his sound poetry, entitled "KLINKT", came out in 2013.

Jan Edler (Germany, 1970) and Tim Edler (Germany, 1965) are the founders of realities:united (realU), a studio for art, architecture and technology. realities:united develops and supports architectural solutions, usually incorporating new media and information technologies. Strategic innovation and a high proportion of communication and mediation in work processes mark many of the firm's projects. Their approach creates a bridge between utopian ideas, abstract conceptions and realization. realities:united works on projects across Europe, Asia and the USA.

Jan van den Berg (The Netherlands, 1961) is best described as an explorer extraordinaire. The expeditions he ventures upon do not so much take him to uncharted spots around the world but to the limits of the naked eye and the naked intellect. He has a penchant for visiting people and (scientific) projects that demand the most of his imaginative capabilities. This provides him with well-informed stories which, when he tells them, soon have the audience wondering whether this is science or fiction. In plain terms: Jan van den Berg is a documentary theatre- and filmmaker in the borderland of performing arts and science.

Jennifer Tipton (USA, 1937) is an influential lighting designer who works in theatre, dance and opera. Tipton also teaches lighting at the Yale School of Drama. She received the Dorothy and Lillian Gish Prize in 2001, the Jerome Robbins Prize in 2003 and in April 2004 the Mayor's Award for Arts and Culture in New York City. In 2008 she was made a United States Artists "Gracie" Fellow and a MacArthur Fellow.

John Day (England, 1969) is the lead insect phylogeneticist at the Centre for Ecology and Hydrology in Wallingford. He trained at the Natural History Museum in the insect genetics of Phlebotomine sand flies before working on bioluminescent beetles for the Ministry of Defence to enhance biological warfare detection systems. He has been fascinated by fireflies and glowworms ever since and has investigated the morphology, taxonomy, population genetics and evolution of these fascinating beetles. He is the Editor-in-Chief of the journal *Lampyrid*, devoted to publishing research on luminescent beetles.

John Pendry (England, 1943) is a theoretical physicist who has made seminal contributions to surface science, disordered systems and photonics. His most recent work has introduced a new class of materials, metamaterials, whose electromagnetic properties depend on their internal structure rather than their chemical constitution. Pendry discovered that a "perfect lens" manufactured from negatively refracting material would circumvent Abbé's diffraction limit to spatial resolution, which has stood for more than a century. His most recent innovation of "transformation optics" gives the metamaterial specifications required to rearrange electromagnetic field configurations at will. In its simplest form the theory shows how we can direct field lines around a given obstacle and thus provide a "cloak of invisibility", which was first realized at Duke in 2006.

Katy Evans-Bush (USA, 1967) is a poet, blogger and freelance writer. Her poetry collections are *Me and the Dead* (2008) and *Egg Printing Explained* (2011). Her blog, Baroque in Hackney, is one of Britain's best-known poetry blogs and was shortlisted in 2012 for the George Orwell Prize for political writing. Her book of essays, *Forgive the Language: Essays on Poetry and Poets*, was published in 2015.

Kobus Kuipers (The Netherlands, 1967) is leader of the NanoOptics group at the FOM-institute for Atomic and Molecular Physics (AMOLF), and member of the Optical Sciences research group. In 2003 he was the recipient of the prestigious NWO-VICI personal subsidy to investigate nonlinear optics at the nanoscale.

Anouk de Hoogh studied molecular life sciences at the University of Wageningen, with a masters specialization in physical chemistry and colloid science. She now investigates optical singularities and nonlinear effects in the near-fields of small nanoscale metallic structures, under supervision of Kobus Kuipers at FOM-institute AMOLF.

Nir Rotenberg studied quantum optics and works at the Max Planck Institute for the Science of Light, in Erlangen, Germany.

Boris le Feber carried out his PhD project on nanoscale electric and magnetic vector fields in the NanoOptics group at the FOM-institute AMOLF. In 2015 he started as a postdoctoral researcher in the group of David Norris at the ETH, Zürich.

Lex Kaper (Netherlands, 1966) is an astronomer whose field of research is massive stars: their formation, evolution and fate. He is the principal investigator of the X-shooter spectrograph on the ESO Very Large Telescope and is currently involved in the development of a multi-object spectrograph for the European Extremely Large Telescope. In January 2016 he was awarded a grant from the National Science Foundation to participate in a study of a multi-object spectrograph, called MOSAIC, to be installed on the E-ELT (2024), which is currently being constructed on the peak of Cerro Armazones near the VLT in Chile.

Marta Brusasca (Italy,1982) studied astrophysics and physics of space at Milan University. She was involved in the design and testing of the first cosine hyperspectral camera and has worked extensively with other optical nondestructive measurement systems. Brusaca's research focuses primarily on food applications of hyperspectral imaging to improve the safety and quality of our daily food.

cosine measurement systems develops and builds measurement systems for its customers. These find use in scientific, industrial, medical, environmental, energy, agri/food, security, semiconductor and space applications, with customers ranging from small high-tech companies to the European Space Agency, IBM and EADS.

Michael Berry (England, 1941) is a theoretical physicist at the University of Bristol who studies mathematical asymptotics to understand connections between physical theories at different levels. He likes to illustrate these abstract ideas with familiar phenomena – the arcane in the mundane – often from optics, with an emphasis on singularities.

Michael Craig-Martin (Ireland, 1941) is an artist who grew up in the US, studying fine art at the Yale School of Art. He has lived and worked in Britain since 1966. Over the past 48 years he has presented numerous solo exhibitions and installations in galleries and museums across the world, including the Centre Pompidou, MoMA, and Kunsthaus Bregenz. Recent solo exhibitions include *Less is still more* at Krefeld Museum Haus Esters; *NOW* at Himalayas Museum, Shanghai and Hubei Museum, Wuhan; and *Transience* at the Serpentine Gallery, London.

Miranda Cheng (Taiwan, 1979) is an assistant professor working at the Institute of Physics and the Korteweg-de Vries Institute of Mathematics of the University of Amsterdam. Her research interests range from mathematical physics, string theory, and various areas in mathematics including geometry, number theory and representation theory. She is in particular obsessed with the possibilities of incorporating new mathematics into string theory, and at the same time creating new mathematics from string theory.

Natalia Zagorska-Thomas (Poland, 1967) is a visual artist who works with *objets trouvés,* often using them in unexpected ways. She has exhibited her work in Australia, Poland, Spain and the UK. She is also a conservator of textiles and textile-related objects, as well as a curator of ExPurgamento, a gallery and salon in Camden Town, London. As a conservator her clients are the National Gallery in London and Buckingham Palace.

David Peggie (England, 1979) is a conservation scientist working at the National Gallery, London. He works closely with conservators and curators, applying a variety of scientific techniques to the characterization of materials in support of conservation treatments and for the understanding of painting technique. His main research interests include the analysis of natural products, such as oils, varnishes and dyestuffs, and the investigation of their deterioration processes.

Oliver Sacks (1933-2015) was a physician, a bestselling author, and a professor of neurology at the NYU School of Medicine. The *New York Times* has referred to him as "the poet laureate of medicine".

Paul Struik (The Netherlands, 1954) is Head of the Centre of Crop Systems Analysis and Professor of Crop Physiology at Wageningen University. His research group focuses on seed systems, crop systems biology, genomics-based crop modelling, multi-scale analysis of photosynthesis, and resource-use efficiency in agriculture. He actively seeks synergy between natural and social sciences through transdisciplinary approaches.

Raihana Ferdous (Bangladesh, 1983) is a PhD researcher at Durham University. Her research interests lie broadly in the field of energy, environment, consumption and development. Currently Ferdous is writing her thesis, which investigates a number of issues connected to the growth of solar energy and electricity consumption in Bangladesh. In 2015 she produced a short documentary, *Off the Grid*, which explores the everyday lives, hopes and dreams of solar energy users in Bangladesh.

Robin Dunbar (England, 1947) is Professor of Evolutionary Psychology at the University of Oxford, and a Fellow of Magdalen College and the British Academy. His principal research interests focus on the evolution of sociality, with particular reference to primates and humans. He is best known for the social brain hypothesis, the gossip theory of language evolution and Dunbar's Number (the limit on the number of relationships that we can manage). His popular science books include *Grooming, Gossip and the Evolution of Language*, *The Human Story*, *Dunbar's Number and Other Evolutionary Quirks* and *The Science of Love and Betrayal.*

Roos van Haaften (The Netherlands, 1983) is an artist with a background in theatre, who moved into large charcoal drawings, and subsequently started to use light and shadow in installation work: "Shadows and silhouettes have my special interest. I arrange objects that, when lit by theatre lights, form shadows or 'drawings' of different tones of grey on the wall. The work is fleeting and temporary, as soon as the lights are dimmed, the image disappears and leaves an empty space." She recently completed a research period at the Institute of Light Design in Amsterdam, where she focused on the subtle complexities of light to make forms and shapes, instead of shadow alone.

Ruxandra Mitache (Romania, 1979) is a visual artist who studied fine arts at Bucharest University of Arts and currently lives in Switzerland. Known for her abstract paintings, Mitache has more recently been exploring themes of the transitive through video and installations.

Siegrun Appelt, (Austria, 1965) is a light artist who since 2011 has been working on *Langsames Licht/Slow Light*, a project that explores light and dark from aesthetic and energy-saving perspectives. The project has also led her to explore through her work new light technologies. Her work is exhibited around the world.

Simon Park (England, 1964) is a Senior Teaching Fellow at the University of Surrey, where he teaches microbiology and molecular biology. He also works at the intersection between art and science and here he has been involved in many innovative art and microbiology projects. The outcomes of these projects have been widely disseminated and they have featured at such venues as the Natural History Museum, the Science Museum and the Wellcome Collection in London.

Steve Northam (England, 1960) spent the early part of his career in production and operations management. Since undertaking a management buyout at defence and aerospace specialist MBM Technology Ltd, Northam has been working on international sales & business development in the defence industry.
Surrey NanoSystems Ltd was founded in 2006 as a spinout from the University of Surrey, and is now a world leader in the development, growth and commercialisation of strategically important nanomaterials, particularly carbon nanotube-based materials. The company's flagship innovation, Vantablack, was developed in 2013 and was measured as the world's darkest material by the National Physical Laboratory in London. Vantablack was publically launched at the Farnborough International Airshow in 2014.

Steven Pinker (USA, 1954) is a cognitive scientist and one of the world's foremost writers on language, mind and human nature. Currently Johnstone Family Professor of Psychology at Harvard University, Pinker has also taught at Stanford and MIT. His research on visual cognition and the psychology of language has won international prizes, as have his numerous books.

Steven Rose's (England, 1954) research is in plasma physics, with a particular emphasis on plasmas produced using high-power lasers. He spends much of his time at the two high-power laser facilities in the UK: the Rutherford Appleton Laboratory's Central Laser Facility, where he became the Associate Director for Physics, and at AWE Aldermaston, where he was the Head of Plasma Physics.

Tadao Ando (Japan, 1941) is an architect with a strong influence on international architectural development. Characteristics of his work include large expanses of unadorned architectural concrete walls combined with wooden or stone floors and large windows. Active natural elements, like sun, rain and wind are a distinctive inclusion to his style. He has designed many notable buildings and he has won many international awards. Ando is honorary member of various architectural institutes around the world. He received the Japanese Order of Culture in 2010.

Tamara Frank (USA, 1957) is Associate Professor at Nova Southeastern University in Dania Beach, Florida, the only state in the continental US that meets her temperature requirements. Much of her research has been on the visual ecology of deepsea animals, studying adaptations to both downwelling light and bioluminescence. She has been chief scientist on 50 research cruises, and participated on 40 more as a lucky hitchhiker.

Tatsuo Miyajima (Japan, 1957) is an artist and Vice President of Tohoku University of Art & Design/ Kyoto University of Art and Design. He first attracted attention in the late 80s for his works using digital LED counters. Since then, his work has been exhibited worldwide, and in international biennales and group shows.

Huang, Xiaoliang (China, 1985) is a photographer who creates magical black-and-white worlds out of projected shadows and layered images. By studying the most fundamental elements of photography – light and shadow – and following the ancient tradition of Chinese shadow puppetry, Huang stretches the boundaries of photographic representation in a masterful display of pre-production skills. He currently divides his time between Changsha and Beijing.

Yoko Seyama (Japan, 1980) is a scenographer and multimedia artist who specializes in spatial time-based art. With a background in architecture and performing arts, her installations combine digital and natural materials, which she works into what she terms a "mutating" space. She has collaborated with world-renowned choreographers and her work has been shown at the Centre for International Light in Unna and light festivals around the world.

Editors

Astrid Alben (The Netherlands, 1973) is a poet and translator who studied English literature and philosophy at Edinburgh University. Alben is the director of PARS and a FRSA, Rijksakademie Fellow, and Wellcome Trust Fellow through the Clore Leadership Programme.

Hester Aardse (The Netherlands, 1971) is a Rijksakademie Fellow whose curiosity and love for beauty in knowledge led her to explore how artists and scientists shape how we see the world. This is expressed in her work: as an urban heritage advisor to the municipality of Amsterdam, and as editor, curator and artistic director of PARS.

Together they are the founders of PARS.

Studio Joost Grootens designs books in the fields of architecture, urban space and art, specializing in atlases, designing both the maps and the books themselves.

Advisors

Astrid Piber, UN Studio
Erika Hoffmann, Hoffmann Collection
Ewine van Dishoeck, Leiden Observatory
Hector Parra, composer
Isabel Nielen, Henk van der Geest,
 ILO Light Research Institute
Jan van Ruitenbeek, Netherlands
 Physical Society, NNV
John Jaspers, Centre for International
 Light Art Unna
Jolanthe Kugler, VITRA design museum
Kobus Kuipers, FOM-institute AMOLF
Rogier van der Heide, Zumtobel
Thomas Hesselberg, Kew Gardens

Credits/sources

In order of appearance:

Cover image: Michael Craig-Martin, artwork © Michael Craig-Martin. Image courtesy of Gagosian Gallery. Photo, Mike Bruce.

François Morellet: published in the catalogue for the exhibition *Kunst Licht Kunst [Art Light Art]*, Van Abbemuseum, Eindhoven, 1966.

Bianca van der Zande: Image 04: K. Sagawa, H. Ujike, T. Sasaki, 2003, "Legibility of Japanese Characters and Sentences as a function of Age", *Proceedings of the IEA 7*: 496–499. Image 06: Van Bommel and van den Beld, 2004.

Tadao Ando: photographs courtesy of the author.

Jan and Tim Edler: text translated by Sheridan Marshall

Katy Evans-Bush: "Poem in which Fashion is a Vacuum" first appeared in *Poems in Which* in 2014.

Iván Navarro: image construction by Thelma Garcia, Iván's studio, Greenpoint Brooklyn, NY, USA. Images installation views by Frank Vinken, International Light Art Award, Unna, Germany.

Marta Brusasca: the hyperspectral camera imaging application to fish freshness was awarded the Herman Wijffels Innovation prize, 2014.

Paul Struik: images courtesy General Electrics Japan

Hamid Ismailov: poem translated by Richard MacKane.

Guy Ropars: image 01: from *The Cultural Atlas of the Viking World,* courtesy of the book editor, James Graham-Campbell.

Tamara Frank: images 16, 19–21 courtesy of *Bioluminescence 2009 Expedition,* NOAA/OER.

Oliver Sacks: "The Case of the Colour-blind Painter" by Oliver Sacks. Copyright © Oliver Sacks, 1996, used by permission of The Wylie Agency (UK) Limited.

Bill Carlton: published with the permission of David Lay Auction House.

Kobus Kuipers et al: we thank C.I. Osorio, T.D. Visser and D. van Oosten for useful discussions. This work is supported by NanoNextNL of the Government of the Netherlands and 130 partners and part of the research programme of the Netherlands Foundation for Fundamental Research on Matter (FOM) and the Netherlands Organization for Scientific Research (NWO), and part of this work has been funded by the project "SPANGL4Q", which acknowledges the financial support of the Future and Emerging Technologies (FET) programme within the Seventh Framework Programme for Research of the European Commission, under FETOpen grant number: FP7–284743. L.K. acknowledges funding from ERC Advanced, Investigator Grant (no. 240438-CONSTANS).

Siegrun Appelt: with the support of Yvonne Ziegler.

Anna Wirz Justice: image 01: M. Terman & S. Fairhurst. Image 02: S. Pierson.

Robin Dunbar: image 01: redrawn from R.I.M. Dunbar, 2014, *Human Evolution,* Pelican, London. Image 02: reproduction with the permission of the Royal Society from E. Pearce & R.I.M. Dunbar, 2012, "Latitudinal variation in light levels drive human visual system size", *Biology Letters*, Vol. 8, 90–93.

Ettore Sottsass: "About Light", *Terrazzo*, 1989, © ADAGP, Paris, 2016.

Steven Pinker: *How the Mind Works*, W.W. Norton, 1997/2009.

Editors
Hester Aardse
Astrid Alben

Copy editor
West Camel

Design
Studio Joost Grootens/
Joost Grootens
Silke Koeck
Dimitri Jeannottat

Lithography
Pieter van der Meer

Printing
Gianotten Printed Media

Paper
Lessebo Design 1.3, 130 grs

Printed in The Netherlands

ISBN 978-3-03778-4907

PARS Foundation
Amsterdam, The Netherlands
www.parsfoundation.com

Lars Müller Publishers
Zürich, Switzerland
www.lars-mueller-publishers.com

Our gratitude goes to the contributors
and advisors, and to the following
people, without whom this book would
not have been possible:
Gert Staal, Hans Mes, Hans Driessen,
Mimi Connell and Clare Tenbeth from
David Lay Auction House, Rombout
Frieling, Menno Liauw, Pjotr van de Jong
and Carla Groen from Vandejong
Creative Agency, Anna Wirz-Justice
and Ken Arnold.

This publication was made possible
with the financial support of:
Stichting Physica
cosine measurement systems
Philips Lighting
montblanc digital productions
Velux Stiftung
Hamasil Stiftung

The editors have tried to trace all
rightful copyright claimants. If there are
persons who we have not been able
to reach, we request them to contact
Lars Müller Publishers.